*The Dynamic Equilibrium Approach*
to Teaching Chemistry

# The Dynamic Equilibrium Approach

## to Teaching Chemistry

Daniel

Luzon

Morris

Parker Publishing Company, Inc.     West Nyack, New York

© 1974, *by*

PARKER PUBLISHING COMPANY, INC.
West Nyack, N.Y.

**Library of Congress Cataloging in Publication Data**

Morris, Daniel Luzon.
  The dynamic equilibrium approach to teaching
chemistry.

  Includes bibliographical references.
  1.  Chemistry--Study and teaching.  2.  Chemistry--
Experiments.  I.  Title.
QD40.M7        540'.7         74-3018
ISBN 0-13-221259-5

## How This Book Will Help You Teach Chemistry

The concept of dynamic equilibrium is one of the unifying concepts of chemistry. It can, and I believe it should, be made one of two or three main threads running through an entire first-year course. This kind of principal theme can change elementary chemistry from the "cookbook" course it sometimes was in the past, to a tightly knit unity, intellectually satisfying in itself. It will also help you provide students with a solid foundation for further study.

You will find here the concept of dynamic equilibrium logically developed, together with some of its more productive applications. The applications in this book include acid-base theory, all ionic reactions, freezing and boiling, reactions of gases—to name only a few—as well as the countless industrial and natural processes that involve these. Indeed, we are just beginning to appreciate the fact that the delicate ecological balance of nature is a dynamic balance, and that our survival as a species may depend on our ability to grasp its laws.

For many years I have taught both standard and advanced chemistry courses. The many activities in this book have been tested in the classroom and used successfully on both levels to supplement standard textbook material. Where many books, for example, give adequate formulas for calculation of freezing and boiling point effects, I strengthen this foundation with a thorough study of the equilibria in liquids, gases, and solids, to make the formulas a simple consequence of these equilibria.

As effective Chemistry teachers, we must be interesting, and when possible, entertaining. For this reason, I start many of my chapters with an interesting, sometimes dramatic demonstration. This is usually something the instructor, or even a student, can do easily. But the entertainment must relate *directly* to the

5

material at hand. You will find, for example, that the "magic" at the start of the chapter on vapor pressure (Chapter 2) is referred to again and again, not only in that chapter, but in many other sections of the book.

First we shall review the gas laws: how they were discovered, and how, in Avogadro's hands, they led to a clear concept of what molecules are. The kinetic theory and van der Waals' laws brought these molecules from abstraction to reality—and by the end of the chapter, you are ready to have your students find the molecular weight of a given liquid.

This leads to the vapor pressure chapter, which serves as a basis for the treatment of equilibrium throughout the rest of the book. Le Chatelier's Principle and collision theory are introduced here, and we go on immediately to a preliminary study of entropy and free energy.

All of these ideas are then applied in a new context. From systems of one component with several phases, we move on to reactions in gases: one phase with many components.

Thus far, the emphasis has been on the conceptual rather than the numerical. Even Gibbs' free energy equation has been used with numbers only to show how the direction of a process can be predicted. But as you present the qualitative properties of solutions, which come next, you get into the quantitative area. If, like me, you dislike the mind-deadening impact of memorized formulas, you'll make use of my "common sense" method for solving molecular weight problems. This, as a method, is useful for many other problems in and outside of chemistry. From here on, where calculations are needed, they will be presented clearly and briefly. With this in mind, I have included a chapter on stoichiometry: calculations from equations, and in titration. The common-sense approach is especially important when we're dealing with the quantitative aspects of buffers, indicators, and hydrolysis. Here, practical calculations rely heavily on the knowledge of what you can and cannot neglect.

The middle section of the book is concerned with ionic equilibria. With solubility products, acid-base theory, and buffers, we lay the foundation for a careful consideration of the Solvay Process for making sodium carbonate. About a dozen ionic reactions must be dealt with in this deceptively simple procedure that "makes the impossible happen." The section winds up with a chapter that offers a unique approach to several analytical procedures.

The chapter on osmotic phenomena touches on areas of great concern to the biologist: colloids the flow of materials through membranes, the nourishment of living cells. We then proceed to electrochemistry, where equilibrium processes clarify some of the basic principles. The final chapter deals with serial processes: the "column" and "tower" processes, which have so greatly extended the capabilities of the technologist in handling both the massive and the tiny. These range from fractional distillation, for the tonnages of the petroleum industry, to chromatography, which can handle nanograms of material for the research man and the toxicologist.

The historical development of a theory is included only when this will help in teaching it. Beyond this I've been arbitrary in my inclusion or omission of historical detail. And I have constantly tried to show the relevance of a principle or process to wider contexts: ion-exchange to soil fertility, vapor pressure to freeze-dried food, acid-base theory to stalactites. This approach can help you break down the ivory-tower image of science and let you show students that science is concerned with the whole real world.

Occasionally, we seem to forget the true foundation stone of science: the belief that *the world makes sense.* Involved with our manipulations and calculations, we sometimes tend to focus on formulas and routines, and neglect their sources. Perhaps this is inevitable. If, walking downstairs, you start worrying about your feet, you may take a tumble. For efficiency's sake, you must almost forget the learning process.

But we are teaching the toddler of science, who isn't really sure where to put his feet, or even why to put them there. To trained scientists (who can go up and down stairs three at a time), some of what I'm presenting here may seem simple. But I have taught long enough to believe that this is really a "working" *science* book. It offers practical suggestions that will enable you to help your students take those first important steps in chemistry.

**Daniel Luzon Morris**

# ACKNOWLEDGMENTS

As you work with the book you will, I hope, find the illustrations useful and sometimes entertaining. Some of them originated in the things I scrawl on the blackboard while I teach. Others were suggested by sketches used in the journal *Chemistry* for some articles of mine. But I am especially grateful to Robert C. Saunders, who took all these and added his own technique and imagination to produce the clean work you see here.

For other acknowledgments: I am everlastingly grateful to the schools where I have taught—The Putney School as a neophyte, and The Lakeside School for the past 20-odd years. In both of these, I was given a free hand to develop the kind of material I present here. Finally, I cannot leave out my thanks to the remarkably competent students of both these schools, who actually taught me a good deal of what I know.

D.L.M.

# Table of Contents

Heating a sealed system until equilibrium vanishes . . .
The disappearance of surface tension, and of heat of vapor-
ization . . . Van der Waals forces in liquid and gas . . .
Liquefaction of gases.

Why things happen: energy loss and entropy gain . . .
Gibbs' free energy . . . A simple free energy calculation . . .
Why ice melts . . . How to get heat from freezing . . .
Dew-point, snowfall, and frozen rutabagas.

Equilibrium between $NO_2$ and $N_2O_4$ . . . Color changes
with temperature . . . Le Chatelier effects . . . Entropy . . .
Reaction rates . . . Effects of pressure . . . The Haber Process
for ammonia . . . Balancing speed against yield.

The ice cream freezer . . . Anti-freezes . . . The liquid
surface . . . Vapor pressure of a solution . . . Raoult's Law . . .
Boiling and freezing points . . . Calculations: common sense
instead of formulas . . . Molecular weight measurement . . .
Non-aqueous solvents.

Electrolytes and their solutions . . . Assumptions of
Faraday and of Arrhenius . . . Conductivity of acetic acid,
water, and their mixture . . . Degree of ionization of acids . . .
From Arrhenius to Debye and Hückel.

Saturated salt solution plus HCl or NaOH . . . The
equilibrium . . . Explanation according to Le Chatelier . . .

Explanation from collisions ... The solubility product ... Puzzles and problems.

Stages in the evolution of a theory ... Weaknesses in Arrhenius' concept of acids ... The ionization of water ... Hydronium ions and the ionization constant ... Solutions of acids ... The pH scale ... Brønsted-Lowry theory ... Acid-base reactions ... Polyprotic acids ... Hydrolysis ... Liquid ammonia as a solvent ... Lewis acids ... The chemistry of "lime" in the Brønsted-Lowry system ... Experiment in pH measurement.

Reactions of gases ... The general stoichiometric problem ... Solutions: normality and equivalents ... Titration ... Problems, with answers.

Spilled acid and sodium bicarbonate ... Neutralization of acetic acid by sodium hydroxide ... By sodium acetate ... Solutions of weak acids, in detail ... Acid plus salt: buffers ... pH calculations for acids alone ... "Legitimate cheating" ... Calculations for buffers ... How does a buffer "buffer"? ... Stress, collisions, and constants ... Biological buffers ... Indicators ... Hydrolysis, in detail ... Experiment: titration with a pH meter ... Problems.

The "impossible process": soda from lime and salt ... What goes into, and comes out of, the tanks ... A dozen or so equilibrium reactions.

*The Dynamic Equilibrium Approach*
to Teaching Chemistry

**1**

## Teaching the Foundations:

## Weighing the Whirling Molecules

Have you ever thought of weighing a molecule? It is not hard to do, provided you first get the molecule moving fast enough, and far enough away from its neighbors. And *that* is not hard to do. If molecules are in the gaseous state, they follow some rather simple laws, and can be weighed in a very simple experiment, which will be outlined at the end of this chapter.

Other than that, the chapter will be concerned with the basic laws of gases, how they were discovered, and some exceptions to them. It may well be that you already know most or all of these. I suggest, therefore, that you skim through this chapter, stopping only where you find material that you do not already know. We need to be sure that we agree on certain fundamentals.

We shall be very much concerned with the mobility of molecules, their *kinetic* behavior. Knowledge of this begins at least as far back as the time of Robert Boyle, in the middle of the seventeenth century. It was Boyle, of course, who discovered the lawfulness of the compression of gases: that the volume of a fixed sample of a gas decreases in exact proportion to the pressure placed on it, provided the temperature remains constant. In his treatise on this ("New Experiments Physico-Mechanical Touching the Spring of the Air," 1662), Boyle speaks of the volume that the air possessed: "I say possessed, not filled." That is, he was clearly aware that the gas could not be compressed unless there was empty space within it. He suggested two possibilities: that the gas might be in some way comparable to lamb's wool, which springs back after compression, or that it perhaps consisted of moving particles.

More than a century later, Jacques Charles and Joseph Gay-Lussac independently discovered the other half of what we now call the General Gas Law. With rise in temperature there is a uniform expansion of a gas, if the pressure is kept constant. By this time, the second of Boyle's guesses seems to have been pretty generally accepted. Daniel Bernoulli had published, in 1738, a picture of gases as consisting of a tremendous number of particles, all in furious motion.

### The General Gas Law

Combining all these, the General Gas Law seems to have been accepted by the time Dalton propounded his great atomic theory, between 1808 and 1810. The law, known now to every student of elementary science, states that the relationship PV/T remains constant, over large variations of temperature and pressure, for a fixed sample of gas. The units in which pressure and temperature are expressed can be quite arbitrary. For pressure, you can use atmospheres, pounds per square inch, torr (mm of mercury), or what you will—*provided your measurements start from zero.*

That business of starting from zero seems pretty obvious for pressure, until you get confused by what is called "gauge pressure," which calls 1 atmosphere zero. You use gauge pressure when pumping up an automobile tire. But the zero was certainly less obvious when it came to temperature. Even the width of the English Channel still influences what is called zero on household thermometers. Charles and Gay-Lussac's discovery led to what seems to be a *natural* zero of temperature that is independent of the units it is measured in, just as a vacuum is the natural zero of pressure.

If a sample of any gas is cooled, keeping the pressure constant, and its volume is measured at each temperature, the Charles/Gay-

Lussac relationship predicts that a graph relating volume to temperature will be a straight line. See Figure 1-1. The remarkable thing is that no matter how big or small the sample of gas, and (within limits) no matter what gas is used, each such straight line, if extrapolated downward to a zero of volume will hit the temperature axis at nearly the same point: $-273^\circ$ C or $-496^\circ$ F. This is the "natural" zero of temperature, and measurements made from here are called "absolute" temperature. These are used in Gas Law calculations. More precisely stated, then, the General Gas Law says that the volume of a gas is directly proportional to the absolute temperature (no matter what the size of the "degree" you choose) and is inversely proportional to the absolute pressure (no matter what units this is expressed in).

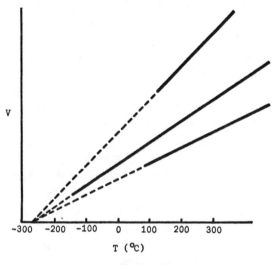

Figure 1-1

Now, the startling thing about the General Gas Law is precisely that it is *general*. It doesn't matter what gas you deal with, air, hydrogen, steam, helium. And it doesn't matter what size sample you take, as long as you stick with the same sample throughout the experiment. All gases show the same pressure-temperature relationships. Or rather they all show *pretty much* the same pressure-temperature relationships. We're going to be concerned from now on with the qualification implied in those words "pretty much."

*Avogadro's Hypothesis*

It seems to have been the generality of the Gas Law that inspired Amadeus Avogadro (in 1814) to propose seriously one of the most outrageous guesses in scientific history. Avogadro believed that all gases behaved alike, under variations in temperature and pressure, because they *were* all alike in one crucial way. This was that under the same conditions, equal containers always contained equal numbers of molecules of gas, no matter what the composition of the gas. Avogadro had a good deal of evidence, other than the gas laws, on which to base his guess, but he definitely did not convince the rest of the scientific community.

Since Avogadro's hypothesis is one of the key building stones of modern chemical theory, we now tend to take it for granted. Yet it is certainly not a likely idea, on its face. Notice what Avogadro says. In the gaseous state, a molecule of hydrogen and one of carbon tetrachloride take up exactly the same amount of space, although the latter is 77 times as heavy as the former. Think of 20 ping-pong balls and 20 bowling balls! It took nearly 50 years for the hypothesis to be accepted: Avogadro himself died before it happened, and it was a student of his, Cannizzaro, who finally convinced the rest of the scientific world. The great chemist Berzelius commented about Cannizzaro's presentation, "It was as though scales had fallen from before my eyes."

By applying this hypothesis, it was now possible to "weigh" molecules of any substance that could be obtained in a gaseous state. All that was needed, for a direct measure of the ratio of the weights of the individual molecules, was to weigh equal volumes of the gases at the same temperature and pressure. Hydrogen, the lightest gas, was assigned an arbitrary value of two units of mass per molecule (the fact that the hydrogen molecule contains two atoms was another consequence of Avogadro's hypothesis), and then all other molecules could be compared to this. Carbon tetrachloride, whose gas is 77 times as dense as hydrogen, would thus have a molecular mass of 154 units.

For convenience, we now use a measure of gases called the "molar volume," which is determined by three arbitrary factors: we use the metric system, and choose "standard temperature and pressure" (abbreviated STP) as 0°C and 1 atm or 760 torr. Under these conditions, 2 g of hydrogen and 71 g of chlorine, for example, both occupy 22.4 liters. The weights of the two gases are their molecular masses, expressed in grams.

Now, under standard conditions, carbon tetrachloride is a *liquid*, and therefore its molar volume cannot be measured directly at STP. However it is easy to measure the mass and volume of some of the material at a higher temperature (say at 100°C), and then use the gas laws to calculate what the volume *would be* at STP. An experiment applying this idea concludes this chapter.

### The Kinetic Theory

So far I have simply taken for granted the assumption of Boyle and Bernoulli, that gases consist of tiny particles in furiously rapid motion. The notion gained gradual acceptance in the first half of the nineteenth century, and was thoroughly sharpened up by Joule in 1848. The kinetic theory of gases postulates that the pressure exerted by a gas is merely the sum of the impacts of its individual molecules on the walls of its container. See Figure 1-2. If each molecule is assumed to be a tiny, perfectly elastic ball, whose kinetic energy is directly proportional to its absolute temperature, then the gas laws can be deduced from this assumption. Not only this, but Brownian movement, the diffusion of gases, and a number of other phenomena, can be easily handled.

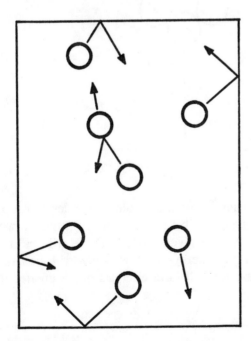

Figure 1-2

One point was recognized early in the development of the kinetic theory. All the molecules of a sample of gas do not have the *same* kinetic energy. They are obviously not all moving in the same direction, and they don't all move at the same speed. At any given instant, some molecules are moving much slower than the average, and some much faster. This idea can be visualized by looking at a case of three molecules, all of the same mass, with two of them moving in on the third. See Figure 1-3. Suppose they all have the same speed to begin with. At the instant of impact much of the energy of both A and B is given to C, which is "squirted" out from between them like a lemon seed from between your fingers. C will now have much higher kinetic energy than it had to begin with, and A and B will have correspondingly less. Yet the average kinetic energy of the three has not changed.

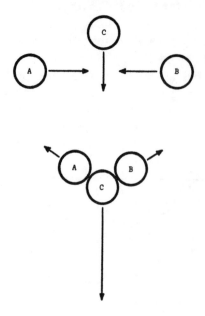

Figure 1-3

The distribution of energies of individual molecules in a typical gas sample can be represented by a curve of the type shown in Figure 1-4. Notice that at a given instant there will be a few molecules with almost no energy—that is, they will be almost stationary—while quite a lot will have energies much greater than the average. The *average* of all these determines the absolute temperature. We'll be very much concerned later with faster and slower members of the population.

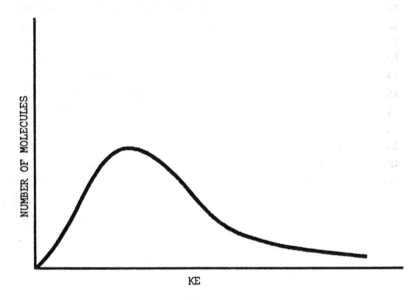

**Figure 1-4**

*Ideal and Real Gases*

The kinetic theory, as much of it as I have presented so far, contains two assumptions that are false: (a) that each molecule in a gas acts as a "dimensionless point"; that is, that its mass has nothing to do with the space it takes up, and (b) that nothing except the walls of its container prevent the expansion of the gas; i.e., that there are no forces attracting one molecule to its neighbors. These assumptions are not bad for gases like hydrogen and helium, but are very far from true for steam or hydrogen chloride. For example, the molar volume of hydrogen (the volume of 2 g of the gas at STP) is 22.43 liters; for hydrogen chloride, it is 22.28 liters. That's a difference of nearly 1%; not good for precision work.

The Dutch physicist, J.D. van der Waals, in 1873, greatly improved the General Gas Law by taking into account the space occupied by the molecules themselves, and the forces between them. The formula that he devised need not concern us here, but his work resulted in the attachment of his name to the forces of attraction between molecules. These forces account for the greatest part of the error in applying the General Gas Law to real gases, and we shall be constantly concerned with the van der Waals forces throughout the rest of this book.

For many practical purposes, it is still convenient to deal with any gas as though it were "ideal"; that is, as though the molecules were spaceless (though not massless) and non-sticky. For such an ideal gas, the molar volume would be 22.414 liters at STP. For a sample of such a gas, if volume is plotted against temperature, the relationship would give a perfectly straight line, which would extrapolate to zero volume at exactly -273°C. See Figure 1-5.

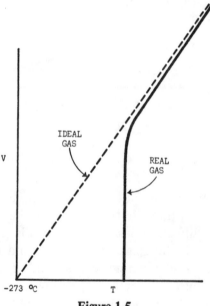

**Figure 1-5**

But every real gas behaves quite differently: at high temperatures and low pressures, a real gas will show a graph that comes close to the ideal. But as this gas is cooled, its volume will begin to fall off perceptibly, below the "ideal" line, and it will finally drop very sharply, as the gas liquefies. The volume of the liquid, while tiny in comparison to that of the gas, is far from being zero. It is approximately equal to the volume of the close-packed molecules themselves, and there is only a small contraction in volume as the liquid is cooled further, until finally it freezes.

In terms of the van der Waals concept, when a substance is in the gaseous state, well above its boiling point, and especially at low pressure, the distance between molecules is great enough so that for practical purposes both the van der Waals forces between them, and their volume, are small enough to be neglected. As the temperature (and pressure) of liquefaction are approached, the attractive forces between molecules begin to have an effect: adding their influence to the walls of the container, they tend to hold the molecules together. The volume of the molecules themselves still plays an almost negligible part in the total volume. Finally the van der Waals forces become strong enough to make the molecules clump, and the gas liquefies.

These then are our assumptions: Gases consist of particles in rapid motion, with relatively great distances between the particles. Their average kinetic energy is directly proportional to the absolute temperature. There are weak forces between neighboring molecules, and the volume of the molecules themselves is very small compared to the space that (to quote Boyle) they "possess, not fill."

*Experiment: Weighing Molecules*

This experiment is a very simple one to perform and gives surprisingly good results.* Appropriate liquids that have given good results in the author's lab are benzene (mol. mass 78), dichloroethane (99) (this is often called "ethylene dichloride"), carbon tetrachloride (154), and ethyl acetate (88). Equipment needed: a flask of 125 or 250 ml capacity, a beaker large enough to contain the flask, a balance accurate to at least 0.01 g, and a balance (it can be cruder) with capacity of 300 g or more.

Method. The flask is weighed, empty, on the accurate balance, together with a square of aluminum foil that will cover its neck. Then a quantity (unweighed) of the liquid sample is poured into the flask (any amount from 2 to 5-8 g). See Figure 1-6. The neck is covered with the foil, which is crimped, and a pinhole is punched in the foil. The flask is immersed as deeply as possible in

*I credit it to Michell J. Sienko, in whose lab manual I first saw it.

gently boiling water. The liquid sample inside presently boils, filling the flask with its vapor (gas), most of which escapes through the pinhole, carrying the air in the flask with it. Heating is continued until the liquid has all boiled away (usually 5-10 minutes). At this point, the flask is exactly full of gas.

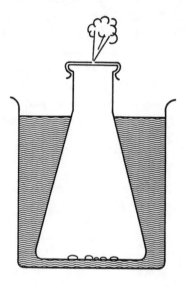

**Figure 1-6**

The flask is now cooled and dried off. As it cools, the gas inside liquefies, but its *weight* doesn't change. The flask is again weighed, and the increase in weight represents the mass of *gas* that was in the flask when it was full, at 100°C. All that remains is to measure the volume of the flask, which is most accurately done by filling it with water and weighing it again (since 1 ml of water weighs 1 g), and we have the data needed for a gas law calculation.

We know the *volume* of the gas; we read the barometer to find *pressure;* its *temperature* is that of the boiling water (and if the altitude of the lab is much above sea level, a correction should be made for the effect of barometric pressure on boiling point). From these three data, we can easily calculate the volume the same gas would occupy at STP if it had not liquefied. Knowing the mass of this quantity of the gas, we find by direct proportion the mass that 22.4 liters would occupy at STP. And this is approximately the mass of 1 mole of the substance.

This experiment may seem crude, yet elementary students repeatedly get values within 5% of the theoretical, using the liquids that I have mentioned (and there are many more that may be used).

It turns out that the principal error in the experiment is caused, not by van der Waals forces (though these are significant), but by the difference between the quantities of *air* in the flask at the initial and final weighing. If a rough correction is made for this,* then normally a molecular weight can be obtained that is only 1% or so *above* the theoretical. This last 1% is presumably caused by the van der Waals forces in the gas as it filled the flask at 100°C.

---

*Correction for air.* It is convenient, and not too inaccurate, to assume that at the time of weighing—i.e., when the vapor has condensed, and air has re-entered the flask—that the vapor pressure of the condensed liquid is about 100 torr. This means that if the barometer stands at 760 torr, 100/760 of the gas in the flask as it sits on the balance pan is vapor of the liquid, while 660/760 of the volume is air. But when the flask was weighed, "empty," at the beginning of the experiment, it was *entirely* filled with air. Thus, 100/760 of the air that was originally weighed with the flask is no longer present, and its weight must be taken into account. The student may take a handbook figure for the density of air at room temperature and apply this to his flask, or work out the weight from the average molecular weight of air (about 29). For a flask of true volume 150 ml (this is about what a 125-ml flask actually holds), the weight of the displaced air is about 20 milligrams. This should be added to the final weight of the experimental flask to compensate for the lost air, and can make a difference of up to 5% in the calculated molecular weight.

The theoretical basis for this correction can be worked out from material presented in Chapter 2.

## 2

Vapor Pressure

and Equilibrium

A surprising number of people have never seen water made to boil by having cold water poured over it—or, for that matter, seen water boil until it freezes. And yet these two occurrences are intimately related to such practical matters as the recipes on the side of a package of cake-mix to be used in Denver, or the preservation of blood serum for emergency use. Mountain climbers know that it is not easy to hard-boil an egg at high altitudes, and a good cook knows that a pressure cooker will greatly speed up the preparation of a meal. Oddities of boiling points explain these things too.

Begin with a demonstration, easy for anybody to do. It requires a flask, preferably a round one, though a small Florence flask will do, with a tight-fitting rubber stopper, not a cork.

Put water in the flask until it is about a quarter full, and start to heat it. See Figure 2-1. When it begins to boil vigorously, put the stopper on top of it in such a way as to let steam escape through a narrow opening. Continue to boil it in this way for a full minute, with steam blowing out past the stopper. This is to blow out practically all the air that was originally in the flask. Now remove the flask from the heat, and at the same time push the stopper in firmly—you have a couple of seconds to do this before the flask cools and lets air back in. Hold the stopper in place for a minute or so. Pressure will then drop inside the flask, and no further holding is needed. Simply make sure the stopper is firmly seated. You now have a rather sophisticated piece of demonstration equipment—or a prop for a magic show, if you prefer.

**Figure 2-1**

First, hold the flask in some way so your fingers don't get burned, and put it under a stream of cold water. It will immediately begin to boil furiously. Take it out of the stream, and the boiling will stop. Put it back, and it will start again. By now the flask will be cold enough so you can easily hold it in your fingers, yet it will still boil hard when cold water hits the top of it.

Just to convince yourself that the world hasn't gone completely topsy-turvy, try warming the flask again (you can do this without risk, as long as it isn't too hot to touch). You'll find that heating makes it boil very satisfactorily too. In fact most of us who have done this demonstration have found that we could go on playing with the same flask for hours, alternately heating and cooling it, and either way make it boil.

There's a third way to make it boil. Cool it to the point where the boiling has practically stopped—yes, this will eventually happen. Then, holding the flask with its neck down, jerk it suddenly downward a foot or so. See Figure 2-2. You'll find that a large bubble forms in the neck, and then it collapses with a clicking sound. This can be repeated indefinitely. The water is in fact boiling in the neck as you move it down, and then the steam recondenses when you stop. The clicking, by the way, is a small-scale replica of the hammering sound that sometimes comes from a steam radiator in winter, and is called a "water hammer."

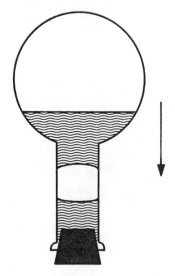

Figure 2-2

To sum up, we have a flask of water that seems to boil whether you heat it, cool it, shake it, or whatever. No matter what you do, it boils. Is this true?

Yes, it is true. You have arranged things so the water in the flask is always exactly at its boiling point. Almost any change that you subject the system to will make the boiling visible. Obviously explanations are in order, which will finally come to what "boiling" means.

*Barometer Tube Experiments*

For these explanations we'll now need more apparatus, somewhat more sophisticated – a pair of barometer tubes. It could well be said that tubes like these started the scientific revolution, in the early part of the seventeenth century, in the hands of Torricelli, Pascal, and Boyle.

Torricelli found that if a long enough tube, closed at one end, is filled with mercury, and then inverted in a cup of mercury without letting any air in, the mercury level in the tube will fall until its surface is roughly 75 cm above the level of the mercury in the cup below (every physics and chemistry book discusses this). In Torricelli's time, there was great controversy over what this meant. Finally, through the efforts of Pascal and Boyle, the scientific world was convinced that there really is a vacuum at the top of the tube, and that the mercury column is held up by the pressure of air at the bottom, rather than by some mysterious, invisible force pulling it at the top.

We are going to use the barometer tube for another purpose than measuring the pressure of the air, but it is important to bear in mind constantly that literally nothing is pushing *down* on top of the mercury, and that air presses steadily upward, always with the same force, no matter what manipulations we are doing.

With a medicine dropper full of ether (acetone, carbon tetra-chloride, or wood alcohol will do fairly well instead of ether), put a few drops of ether into the bottom of one barometer tube, keeping another one as a control. Watch carefully what happens; it will happen only once in a tube (see Figure 2-3). You'll see the liquid ether drop start to trickle up the inside of the tube (because

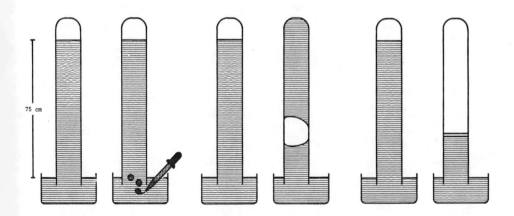

75 cm

Figure 2-3

it is lighter than mercury), until it is about a third of the way up; then it will suddenly start to boil, and will push the entire upper part of the mercury column up the tube, until it hits the closed end with a click. The mercury will then tumble down the tube until it comes to rest with the top of the column somewhere about 25 cm above the level in the cup, with a few drops of ether on top of it. Watch the ether layer: you may see some fine bubbles forming in it—it is in fact constantly boiling!

If you now repeat the addition of ether from below, nothing spectacular happens: the liquid will simply trickle up the inside of the tube to join the layer sitting on top of the mercury. If other liquids are used, the results will be essentially the same. The difference will be in the amount that the mercury level is lowered.

In thinking about this experiment, it is very hard to avoid using old inaccuracies, such as the idea that "suction" was somehow holding the mercury up in the original barometer tube. Until you can get clear of this idea, nothing makes sense. Assume you have.

At the end of the experiment, then, something must be pushing down on the mercury column in the tube with ether. After all, air is still pushing *up* in that tube exactly as hard as it is in the control barometer. You saw the ether boil as it went up the tube. Then it must be the "steam" from the ether that is pushing down.

Here some terms need to be cleared up. It happens that the word *steam* is used only in relation to water; for other liquids, the words "vapor" and "gas" are used. While these last two words are practically interchangeable, in common scientific usage we call a gas a *vapor* only if it is close to its liquid state, by virtue of temperature or pressure. What is usually called steam, the cloud that you see near the spout of a kettle of boiling water, is not vapor or gas: it is a cloud of water droplets. The true steam, water vapor, is the invisible stream of gas issuing from the spout, before it has had time to cool enough to make droplets. All vapors are clear and transparent (though some, like iodine, may be colored).

In our ether barometer tube, then, we have a gas, ether vapor. It presses strongly down on the surface of the mercury, with its pressure opposing the air pressure that is pushing the column up, and at ordinary room temperature the vapor pressure will be roughly two-thirds of an atmosphere. See Figure 2-4. The sum of the vapor pressure, and the pressure of the remaining mercury, will exactly equal the external air pressure. Thus, the vapor pressure of

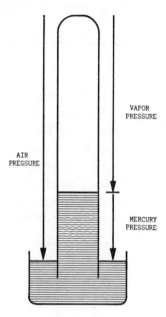

VAPOR
PRESSURE

AIR
PRESSURE

MERCURY
PRESSURE

Figure 2-4

the ether must be equivalent to about 50 cm of mercury or 500 torr.

Now let's investigate the gas in the tube. First, see what effect change in volume has on the pressure. This can be done by tipping the tube slantways, with its base still in the dish of mercury. You find that you can easily push the gas into half its volume in this way. If you now measure the pressure (by measuring the *vertical* height of the mercury column), you will find that it hasn't changed in the slightest. Reduce the volume still more, and the pressure *still* remains the same. In the ideal case, the liquid in the tube could be made to hit the top with a slight click, if it were tipped sideways far enough. And the height of the mercury column at this point—the vertical height, which measures the pressure—will still be unchanged. I said "ideal case" because there is almost always a little air in the tube, that had been dissolved in the ether, and that will not entirely collapse.

Obviously, we have no ordinary gas here. Boyle's Law says that for an enclosed sample of gas, when the volume is halved, the pressure is doubled. With such a gas, reducing the volume to zero would be impossible, requiring infinite pressure. Yet the ether easily goes to half volume, or even to zero volume, with *no* change in pressure.

There is nothing wrong with Boyle's Law. What's wrong with our gas?

Before answering that question, let's see what heat does. Straighten up the tube, and gently warm it by stroking it with a burner flame. See Figure 2-5. Immediately, the pressure increases and the mercury level drops. Long before the tube is too hot to touch, the levels of mercury inside and outside the tube will be equal; that is, the pressure of the gas will reach 1 atmosphere. This seems to violate the Charles/Gay-Lussac Law. To increase the volume of an enclosed gas by one-third, a rise of more than 100°C above room temperature would be required; whereas, here the rise need be only 10-15°C.

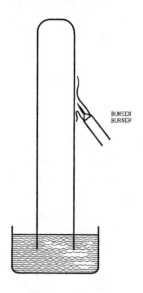

BUNSEN
BURNER

**Figure 2-5**

But careful watching will explain this: as the tube is warmed, you should see the ether layer boiling. Obviously then, we haven't got a "sample of gas" in the sense of the Gas Laws. As the liquid boils, it provides more gas; our sample is constantly changing.

This in turn explains the pressure effect. When we tipped the tube down, trying to increase the pressure, we simply made some of the gas turn into liquid. And when the tipping went far enough, all the gas disappeared. Its volume could become zero because it was all converted into liquid. The Gas Laws simply don't apply. They are true for an enclosed sample of gas *of constant mass*— though this qualification is often overlooked in stating them. Obviously, if you can make more gas any time in the middle of an experiment, or remove some, you're dealing with a totally different ball of wax.

And this was the situation in the sealed flask of the "magic show" with which I started the chapter. That flask was carefully prepared to contain nothing but water and water vapor, just as the barometer tube contains nothing but ether and ether vapor. In both cases, liquid could change into gas, or gas into liquid, with great ease at any temperature.

### The Sealed Flask

The only difference between the barometer experiments and the sealed flask is that the volume of vapor in the sealed flask couldn't change appreciably. When the water in the flask was warmed up by heating it from below (this was done as the second part of the demonstration), it boiled, just as the ether did. In both cases the pressure rose, but since the pressure in the water flask was far below atmospheric pressure, there was no visible evidence · of this. All we were doing was boiling water at low pressure.

How about the cooling, though? Why was there boiling inside the flask when cold water was poured over the outside? Frankly, this was a trick, true legerdemain. The upper part of the flask, containing only vapor, was easier to cool than the lower part, full of warm water (if you had lowered the flask into a pan of cold water, nothing visible would have happened!). Cooling the walls of the upper part of the flask made vapor condense there. Thus some gas was actually removed from action in the flask, and the pressure dropped. Then, with lowered pressure in the flask, the warm water below boiled, to replace the lost vapor pressure.

When the water was made to boil by jerking the flask violently downward, pressure was reduced within the water column; consequently, there could be boiling at some point in the column. As soon as the downward movement ended, the pressure rose and collapsed the vapor to zero volume, with a sharp click.

I have spoken of "boiling" at least six times in the course of the last couple of pages. In fact everything in the sealed flask is explainable in terms of boiling. And boiling is a special case of the *dynamic equilibrium* with which we'll be concerned from here on. To understand both ideas, we must take a hard look at what is believed to happen inside the barometer tube, or the airless water flask.

## Dynamic Equilibrium and Le Chatelier's Principle

I said that the ether in the barometer tube could be seen to be boiling much of the time. We shall assume that in fact it was boiling *all* the time. But since the word "boil" may have special connotations, we say that it was *vaporizing*. We could call it "evaporating," but this word too can be understood in a slightly different sense, so we'll avoid it right now.

It is believed that in a vessel containing nothing but a liquid and its vapor there is constant furious activity going on, even when the surface of the liquid appears to be perfectly still. At every moment a considerable quantity of liquid is breaking loose from the surface and turning to gas; and at the same time, exactly the same number of molecules of gas are returning to the liquid state. The situation is of equilibrium, but the equilibrium is the result of activity, not of things standing still. It is called *dynamic equilibrium*, in contrast to the static equilibrium of two weights hanging at the ends of a balanced lever.

A better example of the contrast is furnished by counting the number of cars in a traffic jam, unable to move (static equilibrium), and an equal number of cars in a crowded shopping center parking lot (dynamic equilibrium). In the latter case, there may always be essentially the same number of cars in the lot, but some are always leaving, to be replaced by newcomers. In a system that is in dynamic equilibrium, there will be very rapid changes if anything alters the rate of one of the two balancing activities. In the parking lot, an hour before closing time, arrivals slack off, while the rate of leaving may still be nearly unchanged, and the lot rapidly empties.

Back to the barometer tube and its ether. Liquid ether is constantly vaporizing, and gaseous ether is constantly condensing. The rate at which the liquid escapes from the surface depends on how fast its molecules are moving when they come to the surface. This depends only on the temperature of the liquid. But the rate at which molecules return to the liquid from the gas depends on two things: how fast they are moving, and how many of them hit the surface in a given time. The rate of return, that is, depends on both temperature and pressure.

Suppose we keep the temperature constant and try to increase the pressure, as we did when we tipped the barometer tube. We saw that we couldn't increase the pressure by decreasing the volume of the vapor. As soon as the gas became even slightly more concentrated, more of its molecules would hit the surface of the liquid and condense. But the rate of escape of molecules from the surface didn't change; therefore, there had to be a net decrease in the number of molecules in the vapor phase. Every attempt to increase the pressure (temperature staying constant) was instantly foiled by loss of the molecules that provided the pressure.

H.L. Le Chatelier proposed a generalization of this kind of effect in 1884. Le Chatelier's Principle, one of the cornerstones of this book, states that: *If a stress is applied to a system in equilibrium, the system will shift in such a way as to decrease the stress.* The example we are considering here is one of the clearest and simplest applications of this Principle. The *stress* in this case was an increase in pressure. The system responded by decreasing the pressure as some of the vapor condensed.

Apply Le Chatelier's Principle to the sealed flask, which boiled when it was heated, cooled, or shaken. The cold water poured on top of the flask made some vapor condense there, producing a stress in the form of decreased pressure. The system responded to this stress by vigorous boiling. Molecules were still escaping at their original rate from the surface of the liquid, but this escape now became visible, since escape was so much more rapid than return. In this case the full original pressure was not restored, because in the process there was some cooling of the whole system. Notice that Le Chatelier's Principle does not say that the applied stress will be eliminated, but only that it will be decreased. The system will have a lower pressure after we have poured cold water over it than it had before, but much higher than it was *during* the pouring of the water.

When we shook the flask and produced the "water hammer" effect, we were again decreasing the pressure near the bottom of the flask by "making a hole" in the water. Immediately the water boiled around that hole, filling it with vapor to bring the pressure up. But as soon as the downward motion of the flask stopped, and pressure of the liquid above was applied to the hole full of vapor, this pressure was decreased as the vapor condensed to zero volume and water hit water with a bang.

Finally, we take up the least mysterious of the cases where the flask boiled: when we heated it from below with a burner, and it boiled hard, though it was still cool enough to hold in the hand. The reason is obvious: in warming the liquid, we increased the speed of its molecules and thus increased the rate at which they could escape from the surface. Since we were warming only the bottom of the flask, we did not (at first) change the rate of condensation of its vapor, so the escape process was very visible. But how did this "decrease the stress" of the applied heat?

In Chapter 1 the point was made that the absolute temperature of a gas measures the *average* kinetic energy of its molecules, and that there is wide variation in the speed of individual molecules. By an easy extension of this idea, we can apply it to a liquid too. And in the state of "furious activity" that I described on page 34 of this chapter where molecules are constantly escaping from the surface and returning to it, the molecules that escape are precisely those of more than average energy. The molecule traveling at a speed near the average will travel through the liquid on a zigzag path, being attracted by all its neighbors uniformly by van der Waals forces, and therefore not being much constrained. See Figure 2-6. But when it reaches the surface it finds no van der

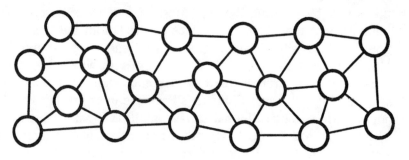

PORTION OF LIQUID NEAR SURFACE.
LINES INDICATE Van der Waals FORCES.

Figure 2-6

Waals forces above it, while those around and below it remain, and it cannot pull loose from these. It takes a molecule of much more than average energy to get away. This means that it is only the "hot" molecules that escape when a liquid vaporizes. See Figure 2-7.

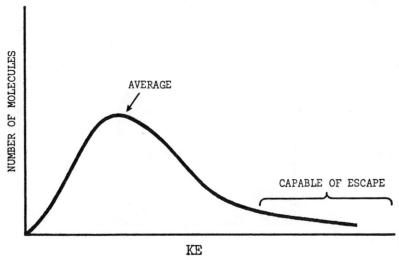

Figure 2-7

So when we warmed the sealed flask, we increased the average speed of its molecules. In boiling, the fastest of these escaped, leaving behind the ones that were still moving slowly. And Le Chatelier was again vindicated: the applied stress—raised tempera-ture—was decreased as the hotter molecules escaped into the vapor phase. In breaking loose from the van der Waals forces, they lost much of their excess kinetic energy, finishing up with a tempera-ture only a little above the average.

Of course, in the process the pressure of the vapor increased, due to both increased speed and concentration of molecules there, so the temperature stress decreased, but was not eliminated altogether.

*The Vapor Pressure Curve*

Without losing sight of the sealed flask and the barometer tubes, which are real pieces of equipment, let's move to an idealized system that we can subject to all kinds of changes, and study in detail.

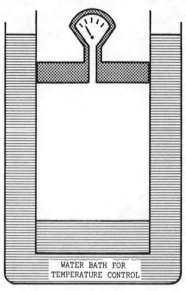

**Figure 2-8**

Imagine a perfectly frictionless piston-and-cylinder arrangement of unlimited length, and provided with a pressure gauge. See Figure 2-8. The gauge reads the true pressure inside the cylinder. The fact that we have atmospheric pressure pushing down on the piston needn't concern us now (as it did with the barometer tube). If the internal pressure is less than an atmosphere, we may have to apply an upward pull on the piston to keep it still, but this pull is against the pressure of the outside air, and has no direct connection with the pressure that shows on the gauge. If we were working on the moon, which has no atmosphere, the gauge reading would be the same, and we'd always have to apply a push on the piston to hold it still. Remember that the ether vapor in the barometer tube actually *forced* the mercury down.

To complete the apparatus, we'll surround the cylinder with a water bath that can be kept at constant temperature by adding or removing heat as required.

The cylinder is provided with a liquid, and its own vapor, but no air or other gas. So the gauge shows the vapor pressure of this liquid.

Now, keeping the temperature constant, move the piston and see what happens. If we move the piston down, the pressure gauge will show a momentary flutter, and then return to just where it was before—until the piston hits the surface of the liquid. At that point, the pressure will suddenly rise. Notice that Le Chatelier's Principle applied only as long as we had an equilibrium: as long as both liquid and vapor were present in the cylinder. The disappearance of vapor ended the condition of equilibrium.

If we move the piston up (still keeping the temperature constant), the gauge will again flutter, this time showing a slight drop in pressure, and then recover as liquid vaporizes. We can go on moving the piston up, with no measurable change in pressure, until . . . . Until again the equilibrium situation is destroyed. This will happen when the last of the liquid disappears. Then the cylinder will contain nothing but vapor—gas. And as a gas, it will obey (approximately) Boyle's Law; that is, as the volume increases with the upward motion of the piston, the pressure gauge will show a drop.

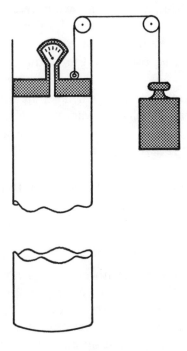

Figure 2-9

Next, return to the original equilibrium situation and try changing the temperature, keeping the *pressure* constant. This may be a little tricky, but not impossible to set up. If the internal pressure is less than atmospheric, we could simply hang a weight as shown in Figure 2-9 to balance the external and internal pressures of the piston. If the internal pressure is more than an atmosphere, of course, a weight can be put directly on the piston. Now try to heat the system. Obviously the liquid will boil, making more vapor, and the piston will rise. But it is the hotter molecules that will vaporize, losing their kinetic energy as they pull free from the van der Waals forces of the liquid.

(Note: We'll look more closely at this loss of energy in Chapter 6—Solutions: Dynamic Equilibrium and the Boiling and Freezing Points.)

The applied heat will therefore be converted into the energy of the gas molecules, and there will be *no increase in temperature*. We can go on adding heat, and letting the piston rise to compensate for it until . . . . Until there is no liquid left. At that point, again, there is no longer an equilibrium: we have a sample of gas, which will obey (approximately) the Charles/Gay-Lussac Law, since we are keeping the pressure constant. If we go on adding heat now, the temperature will rise, and the volume with it, as with any other gas.

Finally, try cooling the system, keeping the pressure constant. Again, no temperature change: vapor condenses, giving up energy in the process, until the piston touches the surface of the liquid. Then equilibrium is destroyed, and cooling the liquid will lower its temperature.

To sum up what we have just been through, the idealized cylinder-piston system is one in which any attempt to change either temperature or pressure, with the other held constant, will fail as long as both liquid and vapor are present.

Next, use the same cylinder-piston apparatus, but this time keep the *volume* constant by anchoring the piston. Now both phases will remain in equilibrium over a wide range of conditions. This time as heat is added to the system, or removed from it, there will be changes in both temperature and pressure. Figure 2-10 shows the type of curve that is always obtained in such an experiment. As the temperature rises, the pressure rises with it, but the relationship is not a linear one (as it would be for a gas alone). Every liquid gives a curve of this type, though the range of values— between hydrogen, say, and mercury or iron—is enormous.

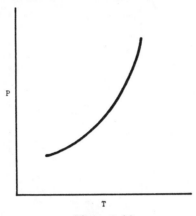

P

T

**Figure 2-10**

*Boiling Points*

Let's see what a vapor pressure curve is good for. On the face of it, few things look less interesting. We'll use a pressure gauge graduated in torr and do an experiment when the barometer stands at 750 torr. We'll use two anchored (constant-volume) cylinder-piston systems side by side, one containing water and its vapor, and the other ether and its vapor. The curves for these are shown in Figure 2-11.

In the first place, it is clear from the graphs that the vapor pressure of ether is very much higher, at any temperature, than is that of water. Now suppose we warm up the ether cylinder until its gauge registers exactly 750 torr—that is, exactly the same as the barometric pressure on the day of the experiment. (See Figure 2-12.) Then the pressure inside the cylinder and outside it are the same, and if we were to drill a hole in the side of the cylinder, nothing at all would happen (except for a little diffusion of air and of vapor through the hole). If we now try to warm up the ether further, in the leaky cylinder, we'll find that it can't be done. The external air acts just like a movable piston, and it moves away as vapor is produced. This maintains constant pressure, letting ether vapor escape, taking away the applied heat. The ether is now at

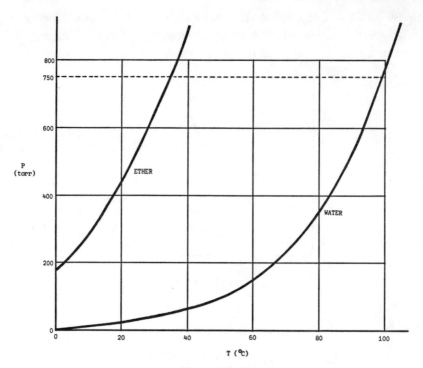

Figure 2-ll. Vapor
pressure curves for ether and water.

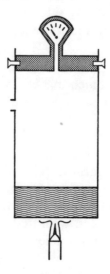

Figure 2-12

what would ordinarily be called its *boiling point* for that day, about 34°C. (Note: The "normal boiling point" found in tables is an artificial one, measured at standard pressure, 760 torr.) At this temperature the vapor pressure of the *water* is only about 50 torr, and if a hole were made in its cylinder, air would rush in, pushed by the external pressure. We'll have to heat the water up to just below 100°C before the pressure gets high enough so we can drill a hole in the cylinder and have nothing happen. The dotted line drawn at the 750 torr pressure, therefore, intersects each vapor pressure curve at the temperature at which that liquid would boil, in an open vessel, on that particular day.

If the same experiment were done on top of Mt. Whitney, where the barometric pressure hovers around 500 torr, the ether would boil (with a hole in the cylinder, or in the open) at about 23°C, and water would boil under the same conditions at about 89°C.

Thus the vapor pressure curve is also a *boiling point curve,* and a good definition for the boiling point of a liquid is something like: *The temperature at which the vapor pressure of the liquid is equal to the pressure of its surroundings.* (Note: This definition is a little more accurate than the one ordinarily used: "The temperature at which the vapor pressure equals atmospheric pressure.") Since in our idealized cylinder piston set-up, the pressure of the surroundings of the liquid *is* its vapor pressure, it follows, as I have repeatedly said, that the liquid in the cylinder is always at its boiling point and the least amount of added heat will always make it boil.

In a household pressure cooker air is expelled, and then the cylinder is kept sealed until the internal pressure is considerably above atmospheric pressure. A common figure is "15 lb," which means 15 lb per square inch *gauge* pressure, or a true pressure of about 2 atm. At this point water boils at about 115°C, and food cooks much more quickly than at 100°C. Hospital autoclaves are large pressure cookers in which things can be perfectly sterilized, with temperatures as high as 125°C.

## Phase Diagrams

We're still not through with the uses of vapor pressure curves. So far we have considered the curve simply as a graphical picture of the temperature and pressure when a liquid and vapor are in equilibrium. That is, we have been interested only in the curved line itself. But the graph is drawn on a two-dimensional surface. Has the rest of the surface any meaning? What is the significance of points like A and B in Figure 2-13?

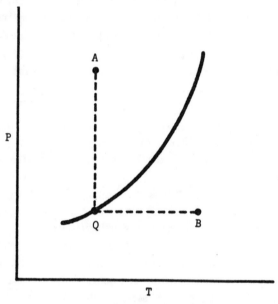

**Figure 2-13**

This question is best answered by experiment. Take a cylinder-piston set-up (no longer anchored) containing liquid and vapor in equilibrium at a temperature and pressure represented by point Q. That is, our pressure gauge and thermometer give those readings. Now suppose we try to change the pressure of the system to the value represented by point A, which has the same temperature as Q but higher pressure. That is we try to "move up" along the vertical line AQ. Look now at the piston (Figure 2-14): in order to increase the pressure, we would have to push the piston down. But we have just seen (page 39) that when we do this, the pressure does *not* increase – pushing the piston down simply makes vapor condense into liquid, with no pressure change. So, although we keep moving the piston downward, the pressure gauge will stubbornly stay at just the value of point Q, as long as the temperature remains constant. How are we going to get to A, then?

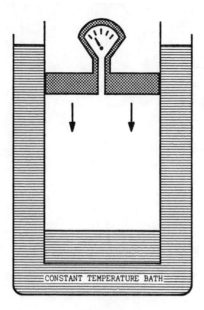

**Figure 2-14**

I've already answered that question. Finally the piston will touch the surface of the liquid, with all the vapor condensed. *Now* the pressure will rise, and rapidly, if we go on pushing down. If nothing but liquid remains in the cylinder, we can easily make the gauge register pressure A, or as much higher as the equipment will stand.

Before considering all the implications of this, let's start from Q and try to reach the conditions represented by B. That is, we must keep the pressure constant, and try to raise the temperature. We have already tried this, on p. 39 of this chapter, and found that it couldn't be done. Again, the conditions remain stubbornly at Q, even though heat is being added. The liquid boils, the volume of vapor increases, the piston moves up—and up—and up. (See Figure 2-15.) The idealized cylinder was supposed to be of indefinite length, so we'll let the piston go on rising, even though the pressure and temperature still are at Q. Finally all the liquid will be gone, boiled away, and we are applying heat to vapor alone. As long as liquid was there, all the heat energy went into making molecules break free of the liquid. But now we can heat the vapor itself, and the added heat *will* speed up the molecules: we can

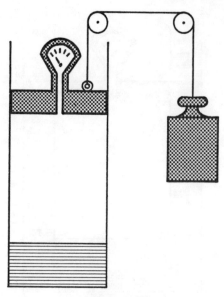

Figure 2-15

easily move to the conditions of B. In heating from Q to B, the
piston will move up, but at a much slower rate than when the
liquid was vaporizing. In fact it will move in accordance with the
Charles/Gay-Lussac Law.

Summing up what happens when we reach those points A and B
on our phase diagram, it appears that we can have nothing but
liquid in the cylinder when the conditions are those of A, and
nothing but vapor when they are those of B. A little thought will
show that we could approach these points by any route we
pleased, without changing the final results. For example, we could
first anchor the piston, and warm the equilibrium system to the
temperature represented by B, while the pressure rose to R. Then
we could remove the anchor, and raise the piston at this new
temperature until all the liquid had vaporized. Finally, raising it
still higher, we could lower the pressure of pure vapor to B. See
Figure 2-16 on following page.

What had started as a vapor pressure curve, and graduated to
being a boiling point curve, has now turned into a *phase diagram.*
See Figure 2-16. A curve of this sort divides the totality of
possible temperature-pressure combinations into two areas. Every
point above the curve represents conditions where liquid alone can

exist; at every point below there can only be vapor. And the dividing line, the curve itself, shows those conditions at which, and only at which, liquid and vapor can exist together.

So far I have dealt with the vapor pressure curve as though it extended upward and downward indefinitely. It does not. Clearly, if you cool a liquid enough, in a fancy cylinder or anywhere else, it is going to freeze, so that gives one limit to the curve. The upper limit is somewhat more complicated, and we'll defer it to the next chapter.

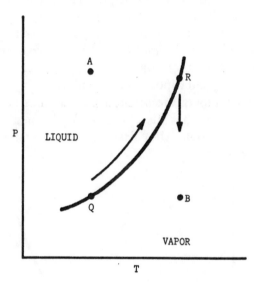

**Figure 2-16**

*Freezing*

Normally when you think of freezing, the idea of melting immediately comes to mind as its opposite. But there are enough freezing-vaporizing cases to make that worth considering also. For example when dry ice, solid carbon dioxide, is exposed to heat under ordinary conditions, it passes directly from the solid to the vapor state. And the frost that forms on grass, trees, and roofs on a cold, clear night has separated directly as solid ice from water vapor in the air.

Since water is the liquid we're most familiar with, we'll put it in an anchored cylinder-and-piston set-up, and investigate its vapor

pressure curve down to, and below, the freezing point. As we cool
it, the pressure will drop in the expected way down to 0°C, and
usually a little below (indicated by the dotted portion of the curve
in Figure 2-17), since water can be supercooled by several degrees
without freezing. If, however, the cylinder is shaken or tapped to
prevent supercooling, the water will begin to freeze when the
temperature is 0°C and the pressure is 5 torr. If more cooling is
now attempted—i.e., more heat is removed—we have another
application of Le Chatelier's Principle. There will be no drop in
temperature, and no change in vapor pressure, until all the water
has frozen. This of course was what happened when we had liquid
and vapor in equilibrium, and tried to change temperature or
pressure, keeping the other constant. The cylinder now has three
phases, liquid, solid, and vapor, all in equilibrium; and any attempt
to change either temperature or pressure will have no effect until
one phase or another has disappeared. We'll come back to this
"triple point" in a minute, but first, continue to remove heat and
let all the water freeze.

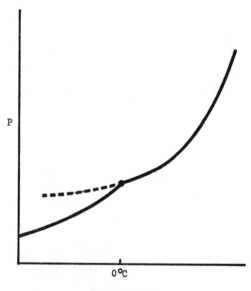

**Figure 2-17**

Once all the liquid has frozen, removal of heat will make the
temperature drop, and the vapor pressure will fall. But the curve
for this is not smoothly continuous with the liquid-vapor curve:

the vapor pressure of the ice drops a little more quickly with decreasing temperature than did that of the supercooled (liquid) water. At $-5°C$ the vapor pressure of ice is about 3.0 torr, while that of supercooled water is 3.1: a small difference, but a very significant one.

An interesting experiment, to show how the shapes of these curves affect the behavior of water, would be the one shown in Figure 2-18. A cylinder-piston set-up is provided with a shelf or cup on one wall. Put some cold water in the bottom of the cylinder and a piece of ice in the cup. Insert the piston and anchor it, put the whole rig in a controlled temperature bath at 0°C, and apply a vacuum pump to remove the air. Presently the water will boil. Let it boil until all air has been swept out, and then seal the cylinder. If we now keep it all at 0°C, the pressure will read 5 torr, and the ice will not melt in its cup, nor will the water freeze. There is complete equilibrium between the three phases, with ice and water vaporizing, and vapor condensing on both.

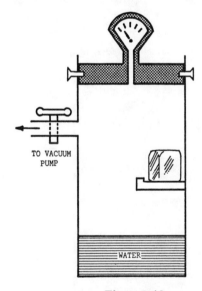

TO VACUUM
PUMP

WATER

Figure 2-18

Very carefully now, cool the whole system until its temperature is $-5°C$. The system is no longer in equilibrium. The pressure of the vapor from the ice should be 3.0 torr, and that of the water 3.1 torr—but vapor from each is the same gas. And vapor molecules are hitting both water and ice at the same rate. Vapor molecules cannot escape from the ice as fast as they do from the water, so the ice will, in a sense, act as a pump, lowering the pressure below the vapor pressure of water. Consequently, the water will boil, rather slowly to be sure, until it has all disappeared. During the process, the pressure gauge will register a pressure *between* 3.0 and 3.1 torr.

The experiment I have just described may be hard to demonstrate in practice: ice crystals will tend to "creep" over the lip of the cup and down into the liquid. As soon as a single crystal meets the liquid, the whole supercooled volume of it will begin to freeze. But the basic idea of the experiment is important: that a liquid, solid, and vapor can exist *in equilibrium* at only one point, one temperature and pressure.

### Liquid-Solid Equilibrium

Our phase diagram is still not quite complete. What happens in a container at the triple point if we try to change the pressure? Again we'll use the cylinder-piston. (See Figure 2-19.) This time it should start with liquid water containing floating chunks of ice,

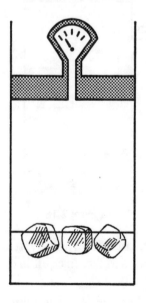

Figure 2-19

and with vapor above, all maintained at the triple point, 0°C and 5 torr. Now move the piston downward, keeping the temperature constant. Obviously vapor will simply condense at first, to form ice and water, and the pressure gauge will show no change. Finally the piston will hit the surface of the water, and the gauge will move up. What will this added pressure do to the ice-water equilibrium?

The results of this experiment may be surprising, and are unpredictable in any given case without experimental evidence. In the case of water, increase in pressure will make the solid ice *melt*. In most other cases of liquid-solid equilibrium, increased pressure will make the liquid freeze. Let's see why.

Look at the ice-water situation from the point of view of Le Chatelier's Principle. This states that when stress is applied to a system in equilibrium, the system will shift in such a way as to decrease the stress. We are increasing the pressure on the liquid-solid mixture maintained at 0°C. The only way this "push" can be decreased is for the volume to decrease. So this happens. When ice melts, it shrinks to about nine-tenths of its original volume. (The opposite of this is the well-known fact that water expands when it freezes, and can burst water pipes.)

Even if, keeping the pressure constant, we had first cooled the ice-water mixture until all the water was frozen, and then had applied extra pressure, the same thing would have happened. With sufficient pressure, ice will melt, even at temperatures well below 0°C. High pressures may be required: roughly 600 atmospheres to make it melt at −5°C. But the effect is highly significant.

Add this information to the phase diagram we have been using for water. See Figure 2-20. I have had to distort all the curves and the scales to make the effect clear. But notice that the water-ice curve slopes to the *left* as the pressure rises; that is, the higher the pressure, the lower the temperature at which ice and water can exist together.

Apply this completed phase diagram to a specific cylinder-piston situation. Begin with a cylinder containing only water *vapor* at −1°C. The pressure of this will have to be very low: let's use 4 torr, represented by point A in the Figure. Now start to raise the pressure, keeping temperature constant. This will be possible at first: as the piston is pushed down, the vapor (gas) becomes more compressed, and conditions move up along the vertical dotted line through A. Presently we reach the ice-vapor curve, and the vapor

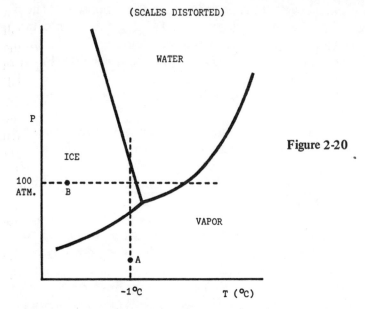

Figure 2-20

begins to freeze, directly from the gaseous to the solid state. We continue to move the piston down, but now the pressure gauge will show no change until all the vapor has frozen.

At this point the piston is sitting solidly on the ice. As we continue to push, the pressure gauge will again rise—granted that we'll need a rather special type of pressure gauge, to show pressure on a pile of snow! But we'll overlook this experimental difficulty. Nothing apparent will happen now until the pressure reaches something over 100 atm. The water will simply decrease very slightly in volume. Then the ice will begin to melt, and its volume will decrease sharply, making the pressure gauge stand still again (even though the piston is moving down) until all the ice is gone. Finally, we'll have water alone in the cylinder. If we reverse the process, that is take the pressure off, the water will at first freeze, and then, if we pull the piston hard enough, vaporize again.

All of that was done at constant temperature: heat had to be removed as the vapor was freezing, and more heat was added as the ice melted, to take care of energy changes in the cylinder. But now suppose that we keep the *pressure* constant, say at 100 atm, and raise the temperature from well below 0°C. The changes in the conditions are shown by following the horizontal line in the diagram, moving to the right from some point such as B.

We'll start with nothing but ice in the cylinder, and the piston in contact with the ice. As we begin to add heat, the solid ice will expand slightly, so the piston must be allowed to rise, to keep the pressure constant. Presently, the melting point will be reached (a little below 0°C) and the ice will begin to melt. As it melts it will decrease considerably in volume, so the piston will move downward to maintain pressure. During the melting process the temperature will stand still, even though heat is being added. When all the ice is gone, the temperature can rise again, and the water will expand slightly. When finally the boiling point is reached (far above 100°C at the pressure we're using), vapor will start to form, and we'll have the now-familiar equilibrium between liquid and vapor. The temperature will stand still, but the piston will rise rapidly as heat is added.

These effects are shown in Figure 2-21 on two related schematic graphs, both having the same time scale. The first shows the motion of the piston, the second the change of temperature.

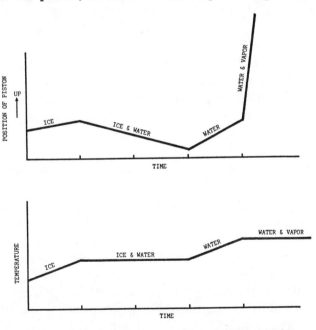

Figure 2-21. Changes in a cylinder which initially contained only ice, as heat is steadily added at constant pressure. Scales are arbitrary, but the time scale is the same for both graphs.

*Practical (and Less Practical) Applications*

So far we have dealt with a highly simplified experimental set-up. Our cylinder-and-piston is not the sort of thing you meet every day on the street, or even in the laboratory. Let's see some of the ways in which the principles we have derived show up in practice.

In the first place, we don't often have water and its vapor in equilibrium with nothing else present. Water standing in a glass or a lake is covered by a layer of air at about 1 atm pressure. Water molecules are still escaping from the surface at the same rate as they were in our artificial systems; so what happens to the equilibrium? It is, in fact, still there. At 22°C the vapor pressure of water, measured in the ways we've already described, is about 20 torr. So just at the surface of the water we would find 760 torr total pressure on the water, 740 torr furnished by molecules of air, and 20 torr by water vapor. On a very moist day, there will be equilibrium exactly like that in our cylinders: if the pressure is increased a little, or the temperature falls, water vapor will condense, and we'll call it fog or rain or dew. If the pressure drops or the temperature rises, water will evaporate faster than it condenses. We'll then have a non-equilibrium system, and water will "boil"; that is, slowly evaporate—slowly, because air molecules will be getting in the way of water molecules as they move into or out of the liquid.

Meteorologists normally express the amount of water in the air in two ways: by the *relative humidity* or the *dew-point*. These figures are two ways of looking at the same information. The relative humidity gives the moisture as a percentage of the amount that would be present if the air were "saturated"; i.e., water vapor in equilibrium with liquid water. The dew-point is the inverse of this: it is the temperature to which the air must be cooled in order to produce liquid-vapor (or solid-vapor) equilibrium. On a wet day, with fog or raindrops in the air to maintain equilibrium, when the temperature is 22°C, the vapor pressure will be 20 torr, the relative humidity 100%, and the dew-point 22°C. If this air is warmed (for example, drawn into a warm room), the vapor pressure and dew-point will be unchanged, but the relative humidity will fall. If the new temperature is 26°C, for example, *saturated* air would contain water at 25 torr, and the relative humidity will be 20/25 or 80%.

What happens if we depart from our vertical and horizontal condition changes on the phase diagram? Suppose we take a liquid and make it evaporate by decreasing the pressure on it, but don't try to keep the temperature constant. In practice, this can be done by simply letting a dry breeze blow on a wet towel. This will remove the vapor—that is, decrease its pressure, as fast as it forms.

Le Chatelier's Principle says that the system must respond by increasing the pressure. This it can do only by evaporating—so it does. But during evaporation the fastest-moving molecules escape, and the remaining liquid becomes colder. This is why you get so cold when you come, wet, out of a swimming pool on a dry day, even though the water and the air are warm.

A surprising application of this effect shows up when a $CO_2$ fire extinguisher is used. See Figure 2-22. This type of extinguisher is a

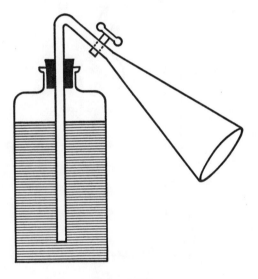

Figure 2-22

heavy tank filled with liquid $CO_2$ in equilibrium with its vapor, exactly like our sealed cylinder-piston rigs. Since the extinguisher has been sitting in the room, its temperature is around 20°C, and the pressure is about 60 atm. We release the valve, and pressure forces the liquid out of the tube. A cloud of white smoke pours out of the horn of the extinguisher, some of it persisting as a white powder on the floor for a few seconds. It is solid $CO_2$ "dry ice." A snowball of this can be collected by putting a cloth over the mouth of the horn. (CAUTION—it will "burn" the skin.)

What has happened is that drops of liquid are forced out of the valve, and each drop instantly begins to boil in the decreased pressure. As I have said, it is the faster-moving molecules that escape, leaving the slower (colder) ones behind, so the temperature of the drop quickly falls. Finally the drop actually freezes as it boils!

"Freeze boiling" used to be a laboratory curiosity, but of recent years it has become important in medicine and industry. The laboratory demonstration of this with water is quite easy to do. A tube of the shape shown in Figure 2-23 is used, containing water and its vapor only. It can be made by applying a vacuum pump just before sealing, or by boiling the water (as we did at the very beginning of this chapter) until all air has been "washed out," and then sealing.

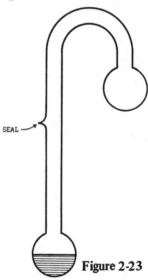

SEAL

Figure 2-23

For the demonstration, all the water is shaken down into the lower bulb, and the upper bulb is placed in a cooling bath. Something like dry ice in acetone is best, but even an ice-salt mixture will work. Immediately the water in the lower bulb starts to boil, as water condenses or freezes in the upper bulb, strongly lowering the pressure. Even when visible boiling slows down, vapor can be seen rippling droplets of water in the ascending tube. Suddenly the liquid in the *lower* bulb freezes, and the demonstrator holds it triumphantly aloft, upside down. Usually the demonstration stops here. But if the upper bulb is left in the cooling bath, something as important, though less spectacular, continues to happen. The ice goes on "boiling." The solid simply vaporizes, exactly as dry ice does under ordinary conditions, until it all disappears.

The first large-scale use of this last effect was by medical scientists, who were trying to preserve blood fractions for longer times than could be achieved by refrigeration alone. This was especially important for emergency field use in war times. See Figure 2-24. It was found that blood plasma (the clear liquid

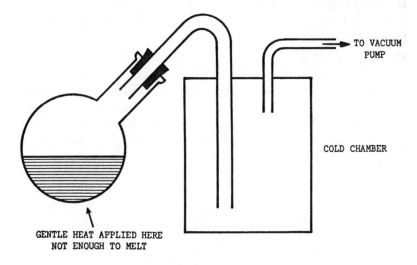

FREEZE-DRYING OF BLOOD PLASMA

**Figure 2-24**

portion of blood, with the cells removed) can be frozen, and then the ice boiled off of the mixture, exactly as it was boiled at the end of the last demonstration. The resulting dry powder (the solids originally dissolved in the water) will keep very well, even at room temperature. Normally the dried plasma is kept in the evacuated container, so its sterility is maintained. When pure, sterile water is added to it, the powder redissolves almost instantly—it had been in a state of almost molecular subdivision— and has the same medical value as the original blood plasma. This technique was developed just before World War II and saved many thousands of lives on the battlefield.

Within the last few years, with the development of highly efficient pumps, the process of "freeze drying" has been applied to many food products, from coffee to vegetables. The process differs only in scale from that used with blood plasma.

## Two Home Experiments

We often warm a vacuum jug before putting hot liquid into it. Sometimes there is a surprise connected with this. Pour about a quarter cup of boiling water into a quart Thermos bottle, quickly put the cap on tight, and swirl or shake the bottle. If the cap is friction-fit, it may pop off; if screw-type, there will be a "pfffft" as it is loosened. Why?

In a well-known demonstration, a piece of fine wire is moved completely through a block of ice without cutting the block. Use a set-up like the one shown in Figure 2-25, with rather fine wire, and a weight of several pounds. Figure out for yourself why the wire will move down but the ice will be solid after the wire has passed through. Le Chatelier's Principle, melting and freezing effects, and heat transfer all come into the explanation.

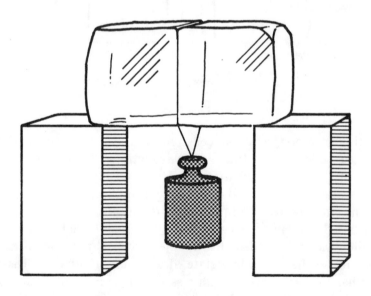

Figure 2-25

# 3

## The Critical Point: Where
## Equilibrium Finally Gives Out

By the time we finished with phase diagrams, in the preceding chapter, it looked as though any phase change we wanted could be accomplished by heating, cooling, compressing, or expanding. Is this always true? To answer this question, we need to follow the vapor pressure curve upward, into the region of very high temperatures and pressures.

For this we should use a sealed vessel, like the one used originally in making the vapor pressure curve. See Figure 3-1. But the vessel needs to be very strong, because we must deal with very high pressure indeed.* Suppose we have such a vessel, and begin to raise the temperature, measuring the pressure as we go.

The liquid will, of course, boil, and the pressure gauge will rise. (See Figure 3-2.) Since we are not letting the piston move, there will be relatively little change in the level of the liquid. The level will drop a little, as liquid boils away, but this will be compensated in part by thermal expansion of the liquid. Finally an extra-

---

*The reader is warned against trying to produce these effects with home-made equipment. I nearly lost a finger once when I tried it!

**Figure 3-1**

ordinary change can be seen in the cylinder: the line that separates liquid from vapor becomes vague—as though a picture was going out of focus—and then disappears. The liquid does not at this point seem to "boil away"; rather, it simply disappears, all at once. As it goes, there may be swirling currents in the vessel, or a sudden opalescence.

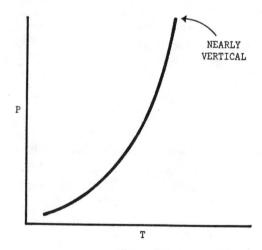

**Figure 3-2**

The temperature at which the change occurs is a definite one, and is called the *critical point*. If the cylinder is heated above this point, the contents behave like a gas (though with very large corrections needed for van der Waals forces).

If the cylinder is now slowly cooled down, from a temperature above its critical point, an even more spectacular process occurs. At the critical temperature, violent swirling is seen inside, followed by what looks like a heavy rainstorm, and the liquid surface reappears, at the same level that it had just before it disappeared. This process, crossing the critical point upward and downward, can be repeated as often as we please.*

If a substance is heated above its critical point, and then compressed (as by pushing a piston down), it is found that no amount of pressure will make the gas liquefy. It will still behave as a gas, though, as I said, with large van der Waals corrections. This was what made early experimenters classify some gases as "permanent"; that is, incapable of being liquefied. We know now that equipment available at that time couldn't produce temperatures below the critical point for gases such as oxygen, nitrogen, hydrogen, etc.

*"The Top of the Curve"*

There is such a sharp change in the behavior of a substance at its critical point that we need to take a new look at our vapor pressure curve and its significance. We found earlier that anywhere on the phase diagram, if the pressure is raised from a point below the curve to a point above it, keeping the temperature constant, we shall first have vapor only; this begins to condense as the pressure value reaches the curve, and finally it all turns to liquid. The pressure on the liquid can now be raised as much as we please. This process is shown by the vertical line AB in Figure 3-3, and will always happen *below* the critical temperature. But if the same thing is tried above the critical point (even barely above it), as shown by line CD, no liquid is formed, no matter how great the value of D. Thus the critical point is an *upper limit* to the vapor pressure curve. Shown as a phase diagram, we should have to draw a vertical line at the critical temperature.

---

*A filmstrip of this, with $SO_2$, is available: "The Critical Temperature of Sulphur Dioxide," BFA Heat Series 435006.

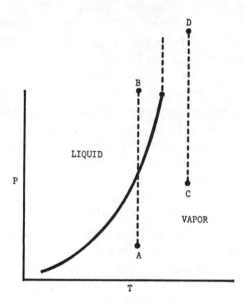

Figure 3-3

*The Molecular View*

To understand what happens at the critical point, we must look again at what happens when a liquid vaporizes. Basically, the picture is quite simple: molecules break free of the surface, overcoming the van der Waals forces which tie them to their neighbors in the liquid. It is only the more energetic molecules that can do this, and they leave less energetic molecules in the liquid. The energy required to restore the loss—to keep the temperature constant when liquid changes to vapor—is measured by the *heat of vaporization:* this is the number of calories required to vaporize 1 gram (or 1 mole) of the material. The value of this heat of vaporization varies with temperature. If we boil water (at low pressure) at 0°C, the heat of vaporization is nearly 600 calories per gram; whereas, at 100°C the value is about 540 calories. This difference would be expected: as the water expands with rise in temperature its molecules are farther apart, and the van der Waals forces are weaker.

As the critical point is approached, the heat of vaporization drops sharply, and at the critical point its value is zero! That is, at this point *no energy* is needed to make a molecule break loose from the liquid.

Before pursuing this concept further, consider the densities of the liquid and the vapor. At 100°C water weighs only a little less than 1 g/cm$^3$. At the same temperature, steam weighs less than a milligram per cm$^3$. But at its critical temperature and pressure (374°C, 218 atm), the density of liquid water has decreased to 0.4 g/cm$^3$, and the density of steam has *increased* to the same value. As the equilibrium system was heated, the water expanded, while the vapor was continually being enriched by the addition of more molecules. At this point then, the molecules of the vapor are spaced exactly as far apart as those of the liquid, and consequently the van der Waals forces are just as strong in one as in the other.

This explains the disappearance of the heat of vaporization. Simply, at the critical point, there is no longer the "surface barrier" at the boundary of the liquid: molecules in the vapor phase, above the surface, can exert as strong a pull upward on an escaping molecule as the downward pull of molecules in the liquid phase, and every molecule is free to move throughout the whole cylinder.

Increase of pressure can no longer convert the vapor to liquid, since the only effect of increased pressure is to increase the forces between molecules all through the vessel.

*Practical Applications*

I have said that early scientists classified gases like oxygen, nitrogen, and hydrogen as "permanent" gases, because they couldn't be liquefied by simple compression. Pressure alone would liquefy chlorine, some oxides of nitrogen, sulfur dioxide, and many others, at room temperature. The difference was merely in their critical temperatures. Those of the "permanent" gases are lower than anything that could be obtained until the latter half of the nineteenth century—nitrogen has a critical temperature of −147°C, oxygen of −119°C. As refrigeration techniques improved, and the significance of the critical temperature was understood, any gas could be liquefied.

For example, in the manufacture of liquid air nowadays, the air is first brought to a pressure of about 100 atmospheres, then cooled by expansion* until the critical temperature is passed, whereupon liquid-vapor equilibrium becomes possible, and the gases can liquefy. See Figure 3-4.

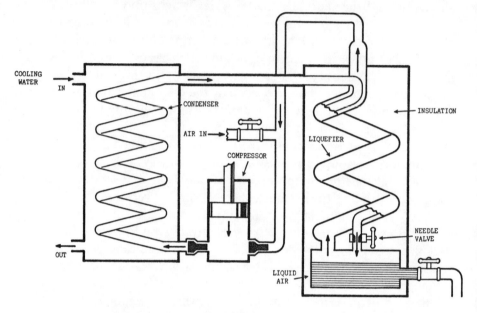

Figure 3-4. A liquid air machine. As air is compressed, it heats up. The compressed air is cooled in the condenser (at left), and led to the inner tube of the double-walled coil at the right. It leaves this tube through the needle valve, and is cooled by expansion there. The cold air, in its return through the outer part of the coil, cools the entering air. Thus the air from the inner tube can become even colder as it expands. This process continues until the air is cold enough to liquefy as it expands (see the discussion of the fire extinguisher, page 55).

A word of caution is necessary at the end of this short treatment. In the preceding chapter we were considering equilibrium situations, and we shall continue to do so throughout the rest of the book. But as we cross the critical point, we *no longer have*

---

*The cooling of a gas by letting it expand depends on the *work* that it must do to overcome van der Waals forces and push away the external pressure. Energy for this work comes from the kinetic energy of the molecules themselves (see next chapter), which are thus slowed to a lower average temperature.

*equilibrium.* It is precisely this that makes the critical point so important: it is the place where equilibrium stops. Above this point we deal with a gas alone, sometimes a gas at very high pressure.

# 4

## Entropy–How to Find
## Direction from Disorder


Everybody knows why a match burns when you strike it, or why a jack-in-the-box pops up when the catch is released, or why a ball rolls downhill. It's easy. Energy had been put into the system at some stage, and when you let it, it shows up again. Most of us can say something about conservation laws, which seems pretty obvious: you can only get out of a system what has been put into it. Energy (or mass-energy, if you're being inclusive) can't be created or destroyed.

This law *seems* obvious now, but it wasn't formally recognized as a general principle until 1842, when Julius Robert Mayer published his famous paper on the subject. The law pulled together a great deal of work that had been done at the turn of the nineteenth century, and it recognized especially the significance of Rumford's experiments on the conversion of mechanical energy to heat (about 1798). It turned out to apply to every kind of energy, from that of a compressed spring, or a magnet, to the heat produced by the metabolism of a living animal. The law makes "perpetual motion," impossible; i.e., a machine that turns out more work than has been put into it. The full and perhaps belated acceptance of the law made the U.S. Patent Office (in 1917) refuse to accept any further applications for patents on perpetual motion devices.

One reason, then, why things happen is because energy has been put into a system, and it can be released again; hence, the jack-in-the-box pops up, or the match burns.

But there are cases where this law—called the *First Law of Thermodynamics*—is insufficient to explain what does, and especially what does not, happen. For example, ice at zero degrees can absorb heat, with no change in temperature, and, in contrast to the compressed spring, it can be very difficult to get that heat back. It will happen spontaneously only under certain limiting conditions. And in an ordinary room there is plenty of heat energy in a glassful of air: its molecules are moving around at a tremendous rate. Yet in order to make use of that heat energy, we must furnish a region of lower temperature, a "heat sink." Then, as "heat flows" from the higher to the lower temperature, work can be done with the kinetic energy of the air molecules in the glass.

There is nothing in the First Law (conservation of energy) that rules out the possibility of separating faster-moving molecules from slower-moving ones. Work could then be gotten out of the resulting pair of systems when heat flowed from the hotter to the colder.

This point is a very subtle one, and I can only suggest its complexities. James Clerk Maxwell first investigated the idea of a tiny energy-less "demon" who might separate fast from slow-moving molecules in order to accomplish what I have just mentioned. He also showed that statistically such a process is simply impossible.

Yet there are circumstances where work can be done by a system that does not change in temperature, or even whose temperature decreases. We worked with such cases rather extensively in Chapter 2. If you merely open the valve of a carbon dioxide fire extinguisher (see page 55), something very definitely does happen. And if you hook up the extinguisher to an engine, you can get quite a lot of work out of it. Yet the final temperature of the system will be well below that of the liquid originally in the extinguisher.

I have said that the first factor that decides whether something will happen is the energy change. There is a most important second factor. It involves the idea of *disorder,* and the Second Law of Thermodynamics formalizes it. In one of the many ways in which it can be stated, it says that *no process can occur unless there is an increase in the disorder of the universe when it happens.* The idea is a very broad one, and when you first meet it, it doesn't seem to make a great deal of sense. It was first used quantitatively by J. Willard Gibbs and furnishes the basis for most modern advances in work-producing machines, from steam engines to nuclear reactors.

The degree of disorder of a system is measured by a function called *entropy.* A restatement of the Second Law would be that in any isolated system (here we needn't consider the whole universe), for any spontaneous process, there is an increase in the entropy of the system. Apply this to the fire extinguisher. We have already seen that when the valve is opened, liquid is forced out and then boils, producing a large volume of gas. If we compare the motion of the molecules in the liquid originally in the fire extinguisher with the gas molecules expanding freely through the space outside, it is fairly obvious that there is a great increase in disorder: each molecule is now free to move through a much greater volume of space. Eventually all the liquid will vaporize, with a tremendous increase in entropy. Even if the final temperature of the gas is much lower than the original temperature of the liquid (and therefore each molecule is moving more slowly), a great deal of work has been done.

We needn't worry about violation of the First Law in this process. Work had been done in converting the carbon dioxide to the liquid state when the extinguisher was filled. This work was available, like the compressedness of a spring, as "potential energy." Yet we have clearly gotten kinetic energy from somewhere. And this somewhere is the entropy, which has increased as the liquid boiled.

That fire extinguisher is a spectacular example of a process in which "energy moves uphill"; i.e., colder molecules drive warmer ones because of increased entropy. But we actually dealt with much simpler cases in our treatment of phase changes.

Consider a cylinder-piston set-up, with water alone in the cylinder to start with. See Figure 4-1. Put a load on the piston, and start to add heat. At first the water will simply warm up, of course, until finally it gets to a temperature where it can boil: where its vapor pressure is equal to the pressure of the air plus the load we have on the piston. Now the piston will start to rise. We can go on adding heat, but there will be no further rise in the temperature in the cylinder, and no further change in pressure. From now on the added heat is doing two things—increasing the entropy of the cylinder contents and doing *work* as the load is lifted. Work is defined as the product of force times the distance through which it moves. This work can be done with no change in (average) speed of the molecules, because of the increase in entropy within the cylinder, as liquid (relatively ordered) becomes gas (relatively disordered). The system is of course a very simple steam engine. It is an inefficient one to be sure: less than 10% of the heat that was put in went into the work of the piston, the rest became "useless" in the entropy increase.

Figure 4-1

This system is perhaps the best one for the introduction of Gibbs' great invention, the *Free Energy* function. This is definitely an invention: nobody has ever seen, heard, or felt of "free energy." But no process can ever occur spontaneously unless the free energy of the system decreases—a match can't light, a jack-in-the-box can't pop, water can't evaporate.

In the next couple of pages, we'll be dealing with quantitative formulas and doing at least one calculation. It should be remembered, though, that the numerical quantities that we'll deal with are not very important, in an introduction to chemistry. What is important, from time to time in the rest of this book, is the *sign* of the quantities. So don't let figures bog you down.

Gibbs' formula is $G = H - TS$, where every symbol except T (absolute temperature) is likely to be unfamiliar. We'll define the new symbols in a minute. The way in which the formula is most often used is for the measurement of *changes*. The standard mathematical symbol $\Delta$ (meaning "difference" or change) is used, and the formula becomes: $\Delta G = \Delta H - \Delta TS$. Now let's look at the meaning of the letters.

$\Delta G$ is the change in free energy, which has the rough significance of energy that can be used to do work. Its exact meaning lies in the fact that in any spontaneous process, it must have a negative value. That is, its sign determines whether such a process can or cannot occur.

$\Delta H$ is the change in heat content of the system, provided the addition or removal of heat is done at constant pressure. Frankly, this term was adopted for the convenience of chemists, who would rather work most of the time with flasks and beakers open to the atmosphere, and therefore at constant pressure. $\Delta H$ used to be called simply the change in "heat content," but variation in pressure may make a lot of difference in the heat added to or removed from a system. So this *constant pressure heat* change is now properly called *enthalpy* change. The word is inconvenient, awkward to pronounce, and confusing. Consequently, many chemists still speak of "heat changes," and you may see tables listing "heat of reaction" when they should properly be given as "enthalpy of reaction." I shall try to use the correct term.

$\Delta TS$ involves changes in two quantities, the absolute temperature and the entropy S. Since we shall be concerned mostly with processes that occur at constant temperature (i.e. no $\Delta T$), this term becomes $T\Delta S$.

Thus, the change in free energy (which decides whether or not something can happen) turns out to be the difference between a *heat* term and an *entropy* term. If much heat is lost (as in the burning of a match), this is likely to be the deciding factor. But if entropy is lost (as in the freezing of ice, where entropy decreases in the orderly formation of crystals), the temperature is the deciding factor.

Come back now to the formula. It is a great deal to swallow in one gulp. We have given a formula by which free energy changes can be measured in terms of entropy, without saying how entropy itself can be measured. But help is on the way.

Suppose we take a system in which there is no change of temperature or pressure when heat is applied. This will neatly fit the needs of the Gibbs formula. And it is, of course, the steam engine just mentioned, or any of our liquid-vapor systems in which the piston is free to move. Any heat added or removed will be $\Delta H$, enthalpy change, since the pressure remains constant; and the temperature will remain constant as liquid vaporizes or condenses. In such a system, $\Delta G$ *will always be zero*, since the system can do nothing spontaneously. True, liquid is always spontaneously vaporizing, but vapor is condensing at precisely the same rate, so the net is no change: $\Delta G = 0$.

This gives us the handle we need. If heat is applied to this system, some liquid will vaporize. Using the Gibbs formula, $0 = \Delta H - T\Delta S$, this becomes: $\Delta H = T\Delta S$. And $\Delta H$ is easy to measure! So finally we can measure entropy.

The job itself is simple arithmetic. For a mole of water at 100°C (373°K) the heat of vaporization (pardon, *enthalpy* of vaporization!) is 9.72 kcal. Therefore, at this temperature the entropy increase when a mole of water boils is $\dfrac{9.72 \text{ kcal}}{373°K}$, or 2.62 kcal per degree. And, by the way, it can be calculated that only about 0.7 kcal of that heat was used to raise the piston.

As I said before, this figure in itself is of no particular significance to us. It was used to show how an entropy change can be calculated, and the basic idea applies to considerably more complicated cases. In fact, any system that is in equilibrium can be used for the same calculation, whether or not any of the applied heat goes into work. Simply, when a system is in equilibrium, its entropy can be increased at the expense of heat energy, or heat can be released as the entropy decreases. We shall be dealing with various kinds of equilibrium from now on, and will refer from time to time to increases or decreases in entropy, but without the need to calculate the size of these changes.

When any substance crystallizes, at its melting point, or from saturated solution, its molecules arrange themselves in a very orderly (low entropy) state. In this process, heat must be evolved:

$\Delta S$ is negative so $\Delta H$ must be negative. There is a nice demonstration of this.

Fill a test tube nearly full of sodium thiosulfate crystals (photographers' "hypo"). Add just enough drops of water to make the crystals look wet half-way up their height. Now warm the tube gently, with steady swirling, over a burner flame. The wet crystals will melt rather readily. When the tube contents are fully liquid, and uniformly enough stirred so no concentration waves show, set the tube aside to cool (if care is used, it can even be cooled in cold water, though in this case premature crystallization may occur). We now have a supercooled liquid. Even though the liquid is far below the freezing point (or saturation temperature), the sodium thiosulfate will not readily crystallize unless some kind of nucleus is provided.

When the tube is cool, drop a single crystal of sodium thiosulfate into the tube—it makes no difference how tiny. Immediately crystals will start spreading from this nucleus, often in beautiful patterns. At the same time the tube will warm up, becoming actually hot to the touch.

This was a non-equilibrium, spontaneous process with negative $\Delta G$. The temperature rose, showing that $\Delta H$ was negative (heat evolved). The entropy became much less as an ordered system was formed, so $\Delta TS$ must have been negative. If the process had occurred under equilibrium conditions (with $\Delta G = 0$ and no temperature change), all the heat would have come from the entropy increase.

This laboratory demonstration has been put to use on a large scale for solar heating. Large tanks of moist "Glauber's salt," $Na_2SO_4 \cdot 10H_2O$, are put in the attic of a specially designed house. The tanks are placed just inside large south-facing windows. During the daytime, heat from the sun is trapped in the attic and melts the salt; at 32°C the crystals break down, and the ions dissolve in the water that is set free. While this is happening, the temperature remains constant at 32°C. All of the sun's heat is used to increase the entropy of the ions, in an equilibrium process for which $\Delta G$ is essentially zero.

When the sun sets, and the surroundings cool down, the salt starts to crystallize (as the thiosulfate did, but without supersaturation, and at constant temperature), setting free heat as entropy decreases. The temperature will still remain at 32°C until all of the salt has crystallized. Pipes carrying circulating water run through the tanks, and to radiators in the house, where they provide steady, mild heat all night.

Heat-vs-entropy effects are intimately involved in many large-scale natural phenomena. See if you can recognize their connection with the following: (a) On a clear evening the temperature will drop steadily until the dew "begins to fall" (it doesn't fall: it forms directly on solid surfaces from water vapor in the air), after which there will be much slower cooling. Weather forecasters take the dew-point into account in predicting minimum night temperatures. (b) There is a folk-saying—true—that when snow begins to fall, the nights won't be so cold. And (c)—on a smaller scale—farmers used to put large buckets of water in root cellars, in the belief—true—that the freezing of the water would in some way "protect" the potatoes and rutabagas from freezing.

**5**

## Teaching Reversible Reactions

## Through Equilibrium in Gases

Anybody who has used nitric acid, and especially the concentrated acid, with metals, is familiar with the heavy, brown fumes that are often formed. These fumes are usually called nitrogen dioxide. The gas is interesting because it is actually a mixture of two kinds of molecules, $NO_2$ and $N_2O_4$, which change very easily, one into the other. In the fumes, therefore, we have a situation similar to the one we dealt with in our liquid-vapor systems, except that here two compounds exist together in one (gaseous) phase, rather than one compound in two phases. The two compounds look different: the simpler, $NO_2$, is brown-colored, and $N_2O_4$ is colorless.

Demonstration equipment for the present discussion is a glass tube containing some of the mixed gases. A test tube will do, though a somewhat longer tube of test tube diameter is a little better. Rubber stoppers can be used at the ends of the tube (though overnight the gases will attack the rubber). The easiest way to make the gas is to heat solid lead nitrate: $Pb(NO_3)_2 \longrightarrow$ $PBO_2 + 2NO_2$. Or copper metal can be warmed with concentrated nitric acid: ($Cu + 4NHO_3 \longrightarrow Cu(NO_3)_2 + 2NO_2 + 2H_2O$. The gas is just heavier enough than air so that a delivery tube can simply be introduced at the bottom of the demonstration tube, and the gas collected by forcing air up and out. See Figure 5-1. There should be a rich brown color in the tube, but no attempt need be made to have the gas free of air.

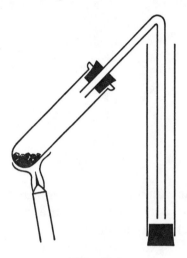

**Figure 5-1**

If now the tube is stoppered at the top, and warmed gently over a burner flame, the color will get darker. If it is cooled by holding it under a stream of cold water, it will turn much paler. If the upper end is warmed, and the lower end cooled, there will be a marked difference in color between the two ends. And if the tube is now inverted, so the lower end is the warmer one, brown swirls will move up from the bottom, to lose their color at the cool top. See Figure 5-2. This action will continue until the temperature at the two ends evens out.

Knowing that the brown gas is $NO_2$, it's easy enough to figure out what must be happening. Heat must be causing $N_2O_4$ molecules to break up into pairs of $NO_2$ molecules, while cooling permits the recombination. Avogadro's Principle predicts that the $N_2O_4$ should be just twice as dense as the $NO_2$, and will therefore sink toward the bottom of the tube if it is formed at the top. (Note: While many reactions of this general type are known, this particular reaction, is one of a very small number that occurs readily, and visibly, at room temperature.)

Figure 5-2

If the equilibrium situation here is similar to those in liquid-vapor systems, we may start with the assumption that two processes are constantly going on:

$$2NO_2 \longrightarrow N_2O_4, \text{ and } N_2O_4 \longrightarrow 2NO_2.$$ And when equilibrium exists, the two reactions are proceeding at the same rate. In other words, we again have *dynamic* equilibrium.

## Le Chatelier

Now let's analyze from several points of view the changes produced by heating or cooling. In the first place we have Le Chatelier's Principle, the oldest, and in many ways the simplest way to look at an equilibrium system. The Principle says that when stress is applied, the system must change in such a way as to reduce the stress. The stress applied was the addition of heat, and brown gas was produced. Then the change from the colorless to the brown gas must be the one that tends to remove—i.e., absorb—heat, and the equation for the reaction is often written:

$$N_2O_4 + heat \longrightarrow 2NO_2, \text{ or } 2NO_2 \longrightarrow N_2O_4 + heat.$$ (Notice that the "heat" term in these chemical equations is not used in balancing.) The second of this pair of equations would be used to predict that you would get more $N_2O_4$ (colorless) by cooling, since the system would respond to the stress of heat removal by providing more heat.

The form of equation that is most often used may tend to be confusing when it is first met:

$$2NO_2 \rightleftharpoons N_2O_4; \Delta H = -13.57 \text{ kcal.}$$ The confusion is contained in the fact that where we had written "+ heat," we now show a negative sign for the heat change. But this sign means that the *system* loses heat when the reaction proceeds to the right *as written*. How it is written is an arbitrary matter. It could equally be:

$$N_2O_4 \rightleftharpoons 2NO_2; \Delta H = +13.57 \text{ kcal.}$$ This would mean that heat must be added to the system when $NO_2$ is formed.

## Entropy

We can also consider entropy changes in the tube with the nitrogen oxides. For a given number of atoms of oxygen and nitrogen, $N_2O_4$ represents a more orderly arrangement than $NO_2$: there are half as many molecules of the first as of the second, and therefore much less confusion as they dart about. This conclusion is confirmed by the use of the equation we met in the entropy chapter:

$$\Delta G = \Delta H - T\Delta S.$$

Since the gases are in equilibrium with each other at any given temperature, $\Delta G = 0$, and $\Delta H$ must equal $T\Delta S$. For the change from $NO_2$ to $N_2O_4$, where heat is evolved, or $\Delta H$ is negative, the $\Delta S$ must also be negative; that is, the entropy decreases.

## Reaction Rates

A third way of considering the reactions in the tube leads us into a new area which will be of considerable importance throughout the rest of this book—the matter of rates of reaction.

Consider an imaginary situation where we had a sample of pure colorless $N_2O_4$. The situation is imaginary because even while a vessel was being filled, the molecules would start breaking down to form brown $NO_2$. But at the hypothetical start of this process, the *rate* of breakdown would obviously depend on how many molecules of $N_2O_4$ were present, and therefore how many moles. The complication, of course, is that as soon as the process has started, the number of moles decreases and the rate falls off.

If we started, on the other hand, with pure $NO_2$, there would again be a hypothetical initial rate of formation of $N_2O_4$, and this rate would depend on the concentration of $NO_2$. For reasons that we'll discuss later (in Chapter 8), the rate in this case is proportional to the *square* of the concentration of $NO_2$. And here also, the rate would fall off as soon as any $NO_2$ was used up.

Clearly, we could start from either end to reach a situation of equilibrium. Suppose we start with pure $N_2O_4$. This begins to break up, and the concentration of $NO_2$ begins to rise. But as soon as any $NO_2$ is present, its molecules will start bumping into each other, to remake $N_2O_4$. See Figure 5-3. At first this is not a very important matter, since very little $NO_2$ is present. But before long $N_2O_4$ is being formed as fast as it is decomposing, especially since less of it is now present than at the start. At this stage, then, the two reactions have achieved exactly equal rates—and we're back to our dynamic equilibrium. Anything that will increase one rate more than the other will shift the relative concentrations of the two gases at equilibrium. We found that higher temperature increased the concentration of $NO_2$, the brown gas. This implies that higher temperature (which is known to increase the speed of *all* reactions) speeds up the decomposition of $N_2O_4$ more than it does the recombination of $NO_2$ molecules.

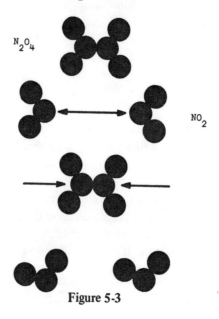

Figure 5-3

## Pressure Effects

We still haven't milked our tube of nitrogen oxides of all the information it contains. What will happen in the tube if we change the pressure on it? This effect is much less easy to demonstrate, so some of what I say here will have to be accepted as reasonable.

Suppose we put some of the equilibrium mixture into a cylinder-piston set-up like the ones we've been using, except that it should be transparent, so we can watch the color. Now push the piston down. Before reading the next paragraph, see if you can figure out how the color will change. Remember, we are applying a stress, increase of pressure. And Le Chatelier tells us that when we do this, the system must respond by a shift in a direction that will decrease the stress.

When $NO_2$ changes to $N_2O_4$, the number of molecules is halved. And Avogadro's Principle says that the volume (or, at constant volume, the pressure) depends on the number of molecules present. Therefore, when $NO_2$ molecules combine, the pressure will decrease. Consequently, when we push the piston down, the color in the cylinder will become *lighter*. (Actually, it would become darker at the instant of pushing, since neither reaction is instantaneous, and *then* become lighter as $NO_2$ is used up.)*

Summing up, if for any reason we wanted a high concentration of $N_2O_4$ in a gas sample, we'd want to cool and compress it; whereas, if we wanted more $NO_2$, we'd do well to heat it and decrease the pressure. It happens that this particular demonstration is of no practical importance, but the principles that have been established here are extremely important for all kinds of commercial and research purposes.

## The Haber Process

An important industrial use of the principles shows up in the Haber Process for making ammonia. It has been known for a long time that heated ammonia gas slowly decomposes into nitrogen

---

*A technicality here. The final color might be a little darker at the end, because we are compressing what $NO_2$ is left, so there may still be somewhat more of it in a *cross-section* of the cylinder than there was at the start. If we looked lengthwise through the tube, there would always be a decrease in color.

and hydrogen: $2NH_3 \longrightarrow N_2 + 3H_2$. This was one of the reactions that was useful in establishing Gay-Lussac's Law of Combining Volumes, which in turn played an important part in Avogadro's considerations. And it was also known that if nitrogen and hydrogen are heated together, they will combine to make very tiny amounts of ammonia gas. Thus, nitrogen, hydrogen, and ammonia form a system like the one we have just used with the oxides of nitrogen, though considerably more complicated, and much slower.

Attempts had been made in the latter part of the nineteenth century to put this reaction to work. Ammonia is important for all kinds of industrial and agricultural purposes. Finally, early in World War I, the process was made fully functional. This happened just in time to give Germany the nitrogenous materials it needed for explosives (via nitric acid) when its supply of Chile saltpeter was cut off by the Allied blockade.

Understanding of the Haber Process involves more detailed consideration of the *rates* of reactions than we've met so far, in addition to factors involving equilibrium. There are three independent conditions that control the rate of any reaction. They are the *temperature*, the *concentration* of each of the materials in contact with one another, and the presence of a *catalyst*. Other conditions are sometimes mentioned, such as physical state, pulverization of solids, etc., but these are merely variants of the three I've mentioned. An increase in temperature will powerfully increase the rate of any reaction. For a rule of thumb, it can be assumed that the rate will be about doubled for every 10°C rise in temperature. Increase in concentration also speeds up rates, but in ways that are not easily predictable except in simple cases. And the presence of specific catalysts will speed up some specific reactions, often by mechanisms not yet understood.

Now, a manufacturer who wants to make something, in this case ammonia, has two main thoughts to concentrate on. How *much* can he make? And how *fast* can he make it?

It was known that when nitrogen and hydrogen combine, heat is evolved. The equation is:

$$N_2 + 3H_2 \rightleftharpoons 2NH_3 \; ; \Delta H = -22 \text{ kcal.}$$ The fact that heat is evolved could have been predicted from entropy considerations. The four molecules originally present will be of higher entropy (disorder) than the two that are formed when they react. There-

fore, there will be *loss* of entropy as the reaction proceeds to the right. Since at equilibrium $\Delta H = T \Delta S$, and $\Delta S$ is negative, $\Delta H$ must also be negative.

This matter of heat is going to be the sticky part of the problem. Obviously, our manufacturer is going to use high pressure. Le Chatelier's Principle dictates this, since the formation of two molecules from four will lower the pressure, and thus relieve the stress. Also, obviously, he'll use a catalyst, for a reaction that is slow at best.

When I say the reaction is slow, I'm understating by several orders of magnitude. You can let nitrogen and hydrogen sit around in a flask at room temperature for years, and still not get enough ammonia to smell. And ammonia gas is used for years in refrigeration plants with no detectable decomposition. We have here a reversible reaction that goes too slowly in *either* direction to be useful at low temperatures, even with high pressures and catalysts.

Obviously, then, our manufacturer must use the third factor—heat. The hotter he makes his gas mixture, the faster it will form ammonia. But that $\Delta H$ figure, which looked rather theoretical when his engineers first mentioned it to him, now raises its head; and an ugly head it is. High temperature will indeed give him ammonia much *faster*, but it will also give him much *less* ammonia at equilibrium. The stress of higher temperature will make the equilibrium shift to relieve the stress—to absorb heat and decompose ammonia!

So the impatient manufacturer is faced with a reaction that doesn't go at all at low temperatures, and that gives poor yields at high temperatures. The only answer is compromise. The Haber Process uses a "moderate" temperature: in the neighborhood of 400°C (cool in comparison with the 1000° to 1500°C that is common in many processes), together with a catalyst (and the nature of the catalysts are sometimes closely guarded secrets). The pressures used are as high as the equipment can stand: 1000 atmospheres or more. Under these conditions, about half the nitrogen originally present is converted to ammonia in a reasonable time.

The manufacturer's tricks aren't exhausted yet. It would be very poor economy for the plant to throw away half the nitrogen and hydrogen that were put into the process. After letting the gases reach equilibrium in the reaction chamber, they are cooled. This liquefies the ammonia, which can be drained off; and the unchanged nitrogen and hydrogen are simply returned to the process. The liquefaction of the ammonia is made possible by the fact that its critical temperature is 132°C, with a vapor pressure at this point ("critical pressure") of only 111 atm. So all that is needed is to cool the gas mixture below this temperature, making use of the pressure already present. See Figure 5-4.

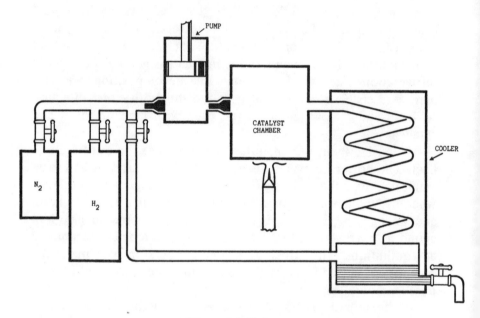

**Figure 5-4. Haber**
**process for making ammonia.**

[Note: If you are reading this book in connection with some other text (and I hope you are), then you may wonder why I don't bring in the *equilibrium constant*. I could, but I would prefer to take it up later, in connection with reactions in solution. You may look ahead there (Chapters 8, 9, and 10) or follow the treatment given in your parallel text.]

# 6

## Dynamic Equilibrium and the
## Boiling and Freezing Points of Solutions

In some of those "good old days," a large part of the population of America knew that the way to make ice cream was to pack custard into a stirring gadget, surround it with ice, add a great deal of salt to the ice, and turn a crank on the gadget. In some way the salt made the ice much colder than normal, and it froze the custard. The result could be very good, and the gadgets are still used occasionally. In those days, ice-cream making furnished the only direct acquaintance most people had with the effect of dissolved substances on the properties of liquids.

Nowadays most of us buy our ice cream ready-made. But with year-round use of anti-freezes a routine part of car care, we have learned that some anti-freezes not only work in winter to keep the radiator from freezing, but also have the rather mysterious effect of keeping it from boiling over in summer. If you take a good "permanent" anti-freeze and mix it with water in a test tube, it's easy enough to measure these effects. If your mixture is of about the same concentration as that used in a car, you'll need some kind of cooling bath to get the temperature low enough to measure the freezing point. Usually a home freezer will do the job, and a mixture of "dry ice" with acetone or alcohol is excellent. As far as the boiling point is concerned, use a bunsen burner, but be sure your thermometer has a high enough range so it doesn't burst its bulb.

The main ingredient in "permanent" anti-freezes now is ethylene glycol, usually containing small amounts of anti-rusting agents, coloring agents, etc. We'll be concerned only with the glycol, and specifically with two aspects of it: (a) Its vapor pressure is low enough to be neglected in all we do with it here—which is why it is called "permanent": it doesn't boil away. (b) A mole of it, as you can easily figure out from its formula, $C_2H_6O_2$, weighs 62 grams.

In Chapter 2, we showed that both the freezing point and the boiling point of a liquid are functions of its vapor pressure. What we need to do now is show what glycol does to the vapor pressure of water. But in order to do that, we need to take a close look at one factor we didn't mention when we put ether into a barometer tube.

Remember what happened there (see Figure 2-3). The liquid ether bubbled up through the mercury until at a certain point in the tube it began to boil. Then the liquid and vapor forced themselves up violently, and we finished with a tube containing about a third as much mercury as we had started with. On this was a layer of liquid ether, and ether vapor occupied the whole top of the tube. See Figure 6-1. In discussing the resulting situation, I talked about the pressure that the vapor exerted on the mercury, completely neglecting the point that the pressure was in fact being exerted on the *liquid ether,* which then transmitted it to the mercury. This point is one I have never seen mentioned in a text, yet it is of prime importance in the picture we are developing here. When a liquid and its vapor are in equilibrium in a container, the vapor exerts pressure, not only on the walls of the container, but also on the liquid itself. This could not happen unless practically all of the molecules of vapor that strike the surface of the liquid bounce off of it, as they bounce off the walls.

We laid stress on the fact that only faster-moving molecules in the liquid have enough energy to be able to escape through the surface. Now we need to take into account the fact that the only *vapor* molecules that can *return* to the liquid are those that hit the surface in the right places. Molecules of liquid are very nearly in contact with one another, and only certain strategic impacts by

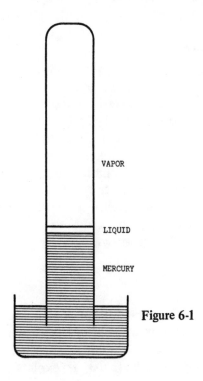

VAPOR

LIQUID

MERCURY

**Figure 6-1**

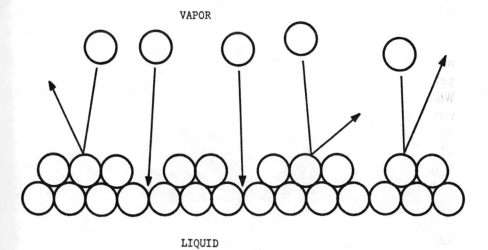

VAPOR

LIQUID

**Figure 6-2**

vapor molecules will be effective—those that happen to hit precisely in a "crack" or in a temporary hole. See Figure 6-2.

This picture clears up a matter that is invariably slurred over in considering liquid-vapor systems—how it happens that the temperature of the liquid and the vapor remain exactly equal. It is clear that the faster-moving molecules of liquid (that is, the hotter ones) that escape from the surface lose some of their kinetic energy as they break loose from the restraining van der Waals forces. But it would be almost magical if the loss of speed sustained by each such molecule were exactly right to bring its kinetic energy back to that of the average in the liquid. What happens, of course, is that the escaping molecule may have any velocity by the time it gets out, from quite fast to very slow. But thereafter, in its repeated elastic impacts with the surface (without penetration), kinetic energy can be exchanged both ways, and the average speed of molecules in both phases will become equal.

## Vapor Pressure of a Solution

Now we are ready to deal with the addition of glycol to water. When we put the mixture into our conventional cylinder-piston device, the pressure gauge will show that the vapor pressure is lower than that of pure water. See Figure 6-3. To explain this, we

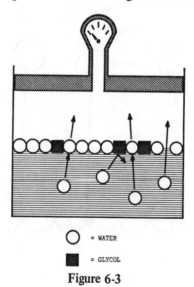

Figure 6-3

need to look separately at the processes of escape and return of water molecules.

For a fast-moving molecule of the liquid to escape, one of two things must happen—either it happens to find a "crack" in the surface, or it hits another water molecule, and knocks that one out. There are very few cracks, as the paragraph above implies, so the second thing is what usually happens. However, if some of the surface is occupied by non-volatile glycol molecules, impacts on these from below won't result in any water vapor. A fast-moving water molecule will simply dissipate its energy after the ineffective bump with glycol—return to the common herd of average-speed molecules. The result is easily predictable: the rate of escape of water molecules will be interfered with in direct proportion to the amount of surface occupied by glycol molecules.

Now, how about the return of vapor molecules to the liquid? In terms of our model, this is simple enough. To return, vapor molecules must find holes or cracks, and it doesn't matter what kind of molecules are at the sides of the hole. See Figure 6-4. We should predict, therefore, that the glycol will have no effect on the return rate.

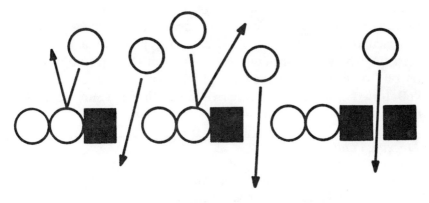

Figure 6-4

If this analysis is correct, the vapor pressure should decrease in direct proportion to the fraction of the surface occupied by glycol molecules. And it does. The law stating this was formulated by Francois-Marie Raoult in 1887 and depends on an extension of Avogadro's Principle. If molecular effects in a *gas* depend only on the number of molecules present, and not on their physical size, the same thing might be true for liquids. As a fairly good approximation, this is true. One mole of any non-volatile material will have the same effect as 1 mole of any other: 62 grams of glycol will lower the vapor pressure as much as 92 grams of glycerol, or 342 grams of cane sugar.*

Raoult's Law states the effect very simply. Instead of calculating the decrease of water vapor pressure due to the glycol, it gives directly the vapor pressure of the water that remains. This is directly proportional to the number of moles of water present in the mixture, called the *mole fraction*, $M_w$. The statement is $P = P_o M_w$, where $P_o$ is the vapor pressure pure water would have at the temperature being used. Thus if we added 1 mole of glycol to 4 moles of water, the mole fraction of water would be 0.8, and its vapor pressure would be 80% of normal, or 20% less than normal.

## Boiling Point

Now we can work out the effect of this vapor pressure lowering on the boiling and freezing points of the solution. Since the boiling point of any liquid is simply the temperature at which the vapor pressure equals the pressure of the surroundings, all we need to do is make a new vapor pressure curve, based on measurements or calculations, and read off the boiling point at any desired pressure. The graph in Figure 6-5 shows how that would be done on one of those rare days when the barometer stands at exactly 760 torr. Given an accurate curve, any other pressure could be used.

## Freezing Point

The effect of glycol, or other solutes, on the freezing point of water is equally easy to work out, providing we are dealing with dilute solutions, where only the water freezes. In this case the solid phase in contact with the solution will be pure ice, and we'd use the same considerations of *rate of escape* to picture the equilibrium here that we used with the liquid and its vapor on page 87. The escape of water molecules from the pure ice is not

---

*We'll consider the very important exceptions to this molar effect in later chapters.

affected by the presence of glycol; whereas, the glycol does get in the way of water molecules approaching the ice surface. The vapor pressure for the solution is simply extended down past the triple point of the pure liquid, as shown in Figure 6-6, and the new freezing point is given by the intersection of the curve for the solution with that for pure ice.

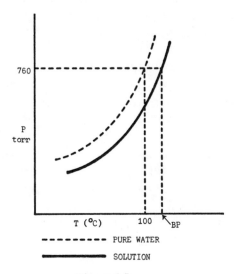

**Figure 6-5**

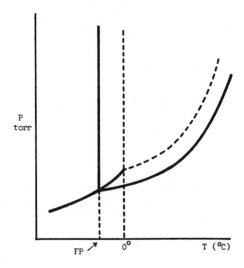

**Figure 6-6**

*Calculation*

The factors that we have taken up here give an adequate picture of what happens, and why, when solutes are added to water. But we need to consider the "how much" in some detail. It would be awkward and time-consuming to make a vapor pressure curve every time we needed to find the boiling point of a solution. So at this point, practical chemists use approximate formulas. Since the vapor pressure lowering depends directly on the number of moles of solute present, and the boiling and freezing point changes depend directly on the vapor pressure, why not eliminate the middleman, the vapor pressure? This is conveniently done by using certain assumptions that are reasonably accurate for dilute solutions (and it is only in dilute solutions that Raoult's Law applies with any precision). The assumptions involve the shape of the curves I've sketched, and needn't concern us here. But it turns out that the raising of the boiling point *above* normal, or the lowering of the freezing point *below* normal, is directly proportional to the number of moles of solute added to a given quantity of water.

The proportionality constants are called the freezing point and boiling point constants. $\Delta FP = 1.86°C \times m$, where $\Delta FP$ is the lowering of the freezing point below 0°C, and m is the number of moles of solute added to a kilogram of water. And $\Delta BP = 0.52°C \times m$, where $\Delta BP$ is the elevation of the boiling point above that of pure water for the prevailing barometric pressure. The term m is called the molality, and differs slightly from molarity, to be discussed later. The difference is minor, and needn't concern us now. Notice that these formulas substitute a given mass of water for the mole fraction of Raoult's Law.

These formulas turned out to be exceedingly important for later theoretical advances, so it is worthwhile to do some calculations with them. We'll work with a series of problems in increasing order of complexity.

*Exercise 1.* What is the boiling point of a 0.2 molal solution of glycerol on a day when pure water boils at 99.2°C?

Solution: The formula gives $\Delta BP = (0.52°) (0.2)$, or 0.10°, so the boiling point would be 99.3°C. The result would be exactly the same for 0.2m solutions of glycol, fructose, cane sugar, or many other substances. (If you have noticed that I've said nothing about salt, I congratulate you. Salts, and certain other types of solutes, give "anomalous" results, which we'll examine in detail in the next chapter.)

*Exercise 2.* What is the freezing point of a solution made by adding 74 g of glycol to 1500 g of water?

Solution: Here we are not given the molality of the solution directly, so we must first discover its value. I have found a scheme like the following to be very useful in *all* problems in which three or more factors must be considered; it takes the place of memorized formulas, and leads to fewer nonsensical errors.

We know that $\left\{\begin{array}{l}\text{1 mole}\\ \text{solute}\end{array}\right.$ added to $\left\{\begin{array}{l}\text{1000 g}\\ \text{water}\end{array}\right.$ gives $\Delta$FP of 1.86°C

We are given $\left\{\begin{array}{l}\text{74 g}\\ \text{solute}\end{array}\right.$ '' $\left\{\begin{array}{l}\text{1500 g}\\ \text{water}\end{array}\right.$ . Suppose we convert

these two figures one at a time.

The concentration was

$$\frac{1000}{1500} \cdot 74\text{ g} \quad '' \quad \left\{\begin{array}{l}\text{1000 g (i.e., we first change}\\ \text{water\quad the weight of water)}\end{array}\right.$$

or          50 g        ''        1000 g

This is          $\dfrac{50}{62}$ moles ''        1000 g (changing from grams
                                                              to moles)

or          0.81 moles   ''        1000 g. Which will give $\Delta$FP =
                                                              (0.81)(1.86°),

then the freezing point will be $-$ 1.51°C.

This method of attack on a problem is cumbersome only the first few times it is used, and is extraordinarily valuable for making "in the head" approximations. We'll apply it in the next two problems.

*Exercise 3.* Recognizing that we'll have errors, since we won't have a dilute solution, about how much glycol should we add to each 0.9 kg (quart) of water to "protect" a car radiator from freezing down to $-20$°C (about $-4$°F)?

Solution: Following the scheme I used in Exercise 2:

1 mole solute added to    1000 g water gives $\Delta$FP =    1.86°

0.9 mole    "       "          900 g   "      "           1.86°

(0.9) (62) g glycol "          900 g   "      "           1.86°

   or 56  g glycol

$\dfrac{20°}{1.86°}$ (56) g glycol "          900 g   "      "           20°

or about 600 grams of glycol, or about 2/3 of a quart.

Experiment shows that this concentration of glycol will give $\Delta FP$ = about 25°.

*Exercise 4.* This presents one of the most important research uses of the freezing point effect—the determination of the molecular weight of an unknown material. A sample of an unidentified sugar weighing 0.300 g is dissolved in 45.0 g of water in a large test tube. The solution is carefully cooled, with stirring, until ice crystals begin to appear. The temperature, read with a thermometer accurate to 0.001°C, is −0.083°. What is the molecular weight of the sugar?

Solution:

0.300  g sugar added to    45 g water gives $\Delta FP$ =    0.083°C

$(\dfrac{1000g}{45g})$ (0.300 g) "    1000 g   "      "           0.083°

   or 6.67 g

Therefore:

$(\dfrac{1.86°}{0.083°})$ (6.67 g) "    1000 g   "      "           1.86°

or 150 g sugar in 1000 g water would give 1.86°.

Therefore the 150 g must be 1 mole of the sugar, and we have found its molecular weight.

*Extension to Other Solvents*

All the factors we have considered in this chapter apply, essentially unchanged, to other solvents. For example, the molal freezing point constant for benzene is 4.9°C, and for naphthalene ("mothballs") is 6.8°C. These figures give the $\Delta$FP when 1 mole of a solute is dissolved in 1000 g of the liquid. The constant for naphthalene is big enough so rough molecular weight measurements can be done with an ordinary laboratory thermometer. Benzene and (melted) naphthalene, moreover, will dissolve many materials that are insoluble in water, so the molecular weights of oils, fats, and many other organic materials can easily be measured.

## Advantages of Freezing Point over Boiling Point

There are three reasons why in the last part of this chapter I have put emphasis on freezing points rather than boiling points. The first two involve practical considerations: (a) the constants for freezing points are always larger than those for boiling points, due to the shape of all vapor pressure curves; and (b) freezing points are not appreciably affected by atmospheric pressure, as boiling points are.

The third reason is more basic. When dealing with freezing points, we need not worry about the volatility of our solute. Since we are dealing only with the exchange of water molecules between the solution and solid ice, we needn't consider the vapor pressure of the dissolved substance. A mole of alcohol (volatile) will give the same freezing point depression as a mole of glycol (nonvolatile).

If we were to measure the *boiling* point of an alcohol solution, we'd have to consider the vaporizing alcohol as well as vaporizing water, and the solution would have a *higher* vapor pressure (therefore lower boiling point) than pure water. This situation can be handled easily, using the principles we've established in this chapter. But we'll defer its consideration to Chapter 16, in connection with fractional distillation.

# 7

## Electrolytes: Dynamic
## Equilibrium and Ionization

I began the last chapter with mention of the ice-and-salt mixture of the old-fashioned ice cream freezer. Yet in the rest of the chapter, I rather carefully avoided further mention of salt as a freezing point depressant. This was because it needs rather special treatment of its own.

The story on this goes back to the 1830's, with Michael Faraday's brilliant work on the conduction of electricity in liquids. Faraday found that water solutions of three classes of chemicals will conduct electricity. These are acids, bases, and salts. Definitions of the first two of these have undergone considerable change since Faraday's time, but the classifications are still useful. The name "electrolyte" is used for these substances. The word comes from Faraday's concept of what happens in the conduction process. He noted that materials in solution were apparently "torn apart" by the electricity, and therefore called the process *electrolysis*; *lysis* is the Greek word for dissolving, or pulling apart.

If a salt like copper chloride is dissolved in water, and direct current is passed through it, copper appears at the cathode (the "negative pole"), and chlorine gas at the anode ("positive"). For this to happen, and go on happening, atoms of the two elements must travel through the solution, and Faraday named these traveling atoms *ions,* which is the Greek word for travelers.

In unrelated work during the next 50 years, much was being learned about solutions, and the effects of dissolved substances on freezing and boiling point were worked out. These were molar effects, as we saw in the last chapter, and their laws were simple. They worked well for practically any solvent and practically any solute—with the notable exception of the most common solvent of all, water, if the solute was an electrolyte.

If you dissolved sugar (a non-electrolyte) in water, or aluminum chloride (an electrolyte) in dibromoethane, the freezing point effects were predictable on the basis of the number of moles used. But if the aluminum chloride was dissolved in water, the freezing point was depressed much more than would be predicted by the standard formula—three to four times as much.

The answer came from a young doctoral candidate named Svante Arrhenius, in 1884—very nearly at the cost of his doctorate. (His thesis was given the lowest possible "passing" grade.) He suggested that the "anomalies" that resulted when electrolytes were dissolved in water could be explained if it were assumed that Faraday's ions were actually caused by the water itself; that the electric current had nothing to do with their creation. As soon as copper chloride or hydrogen chloride or potassium hydroxide (these three represent a salt, an acid, and a base) dissolve in water, the water goes to work on their molecules, partially breaking them apart into ions. Note that the word "partially" would be Arrhenius', and is accepted today only for acids. We'll follow through with Arrhenius' explanation for the case where present ideas are close to his.

He would say, then, that when hydrogen chloride is dissolved in water, it *immediately* ionizes, though not completely. The solution still contains hydrogen chloride molecules, but it is mostly a solution of hydrogen ions and chloride ions: $HCl \rightleftharpoons H^+ + Cl^-$. When electricity is passed through the solution, these ions will travel exactly as Faraday supposed, the positive ones to the negative pole, and the negative ones to the positive. *But each ion will have the same effect on boiling or freezing point as a molecule would have.* Thus a 1 molal solution of hydrochloric acid would give a freezing point depression of something less than *twice*

1.86°C; the "something less" being explained by the incomplete ionization of the acid.

Arrhenius' theory, which finally won over even the professors who had nearly rejected his thesis, gave explanations so accurate, and so simple, that we still use it, though we have modified it in many details. We'll work now with the basic ideas.

Some of the most convincing evidence for Arrhenius' model appears in connection with solutions of weak acids—their conductivity and their freezing points. A very simple piece of apparatus for testing the electrical conductivity of solutions is the one shown in Figure 7-1. All that is needed is a ceramic light socket and some wire. With a 25-watt light bulb (preferably clear glass) and test wires 10-15 cm long, you can test any solution, and check what I say here about conductivity. To test a solution, dip the test wires in it, and plug in the cord. The brightness of the light shows the conductivity.*

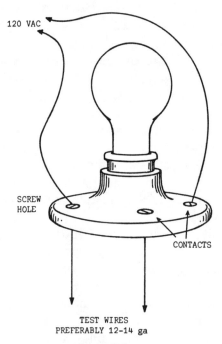

**Figure 7-1**

---

*If you have worked with *resistance* in the past, note that conductivity is simply its reciprocal, and is measured in reciprocal ohms, or mhos.

Acetic acid is a liquid when it is pure (in a reasonably warm laboratory). It is a "weak" acid in the sense that it will do the same things in solution that hydrochloric acid does, but less strongly. In solutions of comparable strength, it will dissolve zinc, or calcium carbonate, more slowly than hydrochloric acid; it will taste much less sour; and it is much less destructive to skin, cloth, or wood. In fact, vinegar is acetic acid solution, a little less than 1 molal in concentration.

Take two small containers, one containing pure water (preferably distilled, unless you live in a "soft water" region) and one with pure acetic acid (usually labeled "glacial"). If you test these two liquids in succession, drying the test wires between tests, you'll find that neither one will light the 25-watt bulb. In fact, either of them makes a good insulator if it is very pure. But if the two are mixed, in any proportions, the bulb promptly lights up. Arrhenius interpreted this as being caused by the ionization of the acid by the water, and so do we.

Now do the same thing somewhat more quantitatively. Make up solutions of acetic acid and hydrochloric acid that are roughly 0.1 molal.* If these two solutions are compared, the difference is startling: the hydrochloric acid solution will light the bulb brilliantly; whereas, with the acetic acid you may barely be able to see the filament glow.

If you have equipment that will allow it, now test the freezing points of comparable solutions in water of acetic acid and hydrochloric acid. You will find that acetic acid gives a $\Delta$FP that is almost exactly what would be expected for sugar or glycol. But with hydrochloric acid, the $\Delta$FP is nearly double this.

With thermometers of high precision, it can be shown that in 1 molal solution the $\Delta$FP for acetic acid is about ½ of 1% lower than would be predicted if exactly 1 mole of molecules were there. This figure can be used for an estimate of the amount of it that has ionized.

Suppose one molecule out of 100 ionizes, giving a hydrogen ion and an acetate ion. The solution will then contain the remaining 99 intact molecules, plus the two ions, or a total of 101 particles from the original 100. These will make the $\Delta$FP just 1% greater than "normal." Obviously then, if the $\Delta$FP effect is ½% greater

*No great precision is needed. Assume that glacial acetic acid contains 17 moles per liter, and concentrated hydrochloric acid about 12.

than normal, one molecule out of 200 has ionized, or one-half of 1%.

With hydrochloric acid, where the $\Delta$FP is nearly twice normal, the ionization has proceeded nearly 100%, in Arrhenius' terms.

And these results very neatly explain the difference in conductivity of the two solutions. In the acetic acid solution, there are only about 1/200 as many ions to carry the current as in the hydrochloric acid.

Considerable food for thought is provided by another simple demonstration with the conductivity-tester of page 96. Two small beakers contain 0.1m solutions of acetic acid and of ammonia, respectively. Each of these is a weak electrolyte, ionized to the same extent. Therefore, each will light the bulb of the tester very feebly. But if the two solutions are poured together and again tested, the bulb will light brightly. This means that the two solutions, containing very few ions, are caused to ionize strongly when they are mixed.

Arrhenius' explanation of this was that when the hydrogen ion of the acid met the hydroxide ion of the ammonia solution, these combined to form water:

$$H^+ + OH^- \longrightarrow H_2O.$$ But this imposed a stress on each of the separate equilibrium situations. For the acetic acid the equilibrium had been:

$$HC_2H_3O_2 \rightleftharpoons H^+ + C_2H_3O_2^-,$$ and for ammonia it had been:

$$NH_4OH \rightleftharpoons NH_4^+ + OH^-.$$

When the solutions mixed, both hydrogen ions and hydroxide ions were removed. As a result (following Le Chatelier's Principle), each equilibrium shifted toward the right to relieve the stress; that is, to increase the concentrations of each of these two ions. Obviously, this set free more ammonium and acetate ions. This equilibrium shift continued until both the acid and the ammonia were completely ionized; and this result was made visible by the brilliance of the light bulb.

When we come to a more complete treatment of acids and bases, in Chapter 9, we shall see that Arrhenius' ideas have had to be modified slightly, with respect both to acids and ammonia. But the basic ideas discussed here remain essentially unchanged by the modifications.

All of these results, which could be extended to water solutions of all kinds of electrolytes, gave a very consistent picture; and Arrhenius' theory of partial ionization held sway for nearly 50 years.

## Ionic Crystals

Changes in the theory came about as a result of deeper understanding of the nature of crystals of ionic materials. For instance, it was found that in sodium chloride (salt) crystals, electrons had already been exchanged between sodium and chlorine atoms. See Figure 7-2. The ions in this case were neither created by the electric current, as Faraday had supposed, nor by being dissolved in water, as Arrhenius postulated. They were already pre-formed in the crystal, and all the water did was dissolve them. These pre-formed ions exist only in salts and hydroxides (Arrhenius' bases); no notable change was needed in Arrhenius' ideas about acids.

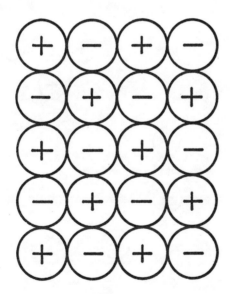

Figure 7-2

This required a new look at the concept of "partial ionization," for the facts were still very much there: sodium chloride solutions did *not* give ΔFP values that corresponded to the complete ionization that was known to exist, nor was their conductivity what would be expected for this. The term "activity" was substituted for "concentration," and it was clear that as solutions of salts became more concentrated, the "activity" of their ions decreased. Arrhenius had said that they were less fully ionized, and this was just the way they behaved.

### The Debye-Hückel Theory

The anomaly was cleared up by Peter Debye and E. Hückel in 1923. Their explanation is too complicated to consider in detail here, but in its basics it is simple enough. When sodium chloride dissolves in water, the pre-formed ions do separate, acting as individual particles. But there are strong electrical forces between these dissolved ions still, though they are surrounded and separated by water molecules. As a result of these forces, their freedom of motion is hampered. They behave as though they were

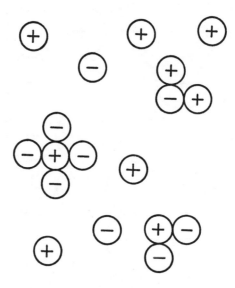

Figure 7-3

forming clumps, if only temporarily, and each clump, for the duration of its existence, behaves like a single particle. See Figure 7-3. The net result, over the whole volume of the liquid, is a smaller *effective* number of particles. This is what shows up, in both the vapor-pressure-related effects, and in the conductivity, as lower "ionization."

The more dilute the solution, the less the probability of this "clumping"; and for the calculations that we shall use throughout this book, we'll continue to work in terms of concentration, rather than with activity, ionic strength, and other important refinements that are needed for precise work.

Finally, the Debye-Hückel effect applies also to strong acids, like hydrochloric. These, like salts, are well over 99% ionized in water solution (though not in their molecular state; Arrhenius was completely right there). But they don't act that way when the number of particles is measured by either the $\Delta FP$ or the electrical conductivity. Their ions also tend to clump.

## Summary

Summing up this chapter, we have seen that the behavior of water solutions of certain materials, both in relation to the electric current and the effects that are related to vapor pressure, led to a concept of ionization that underwent gradual change from Faraday's time to the present.

For Faraday, *electricity* broke up the molecules of acids, bases, and salts into their ions, which could then travel under the influence of electric charges.

For Arrhenius, *water* broke up these molecules partially, and the ions so formed not only traveled, as Faraday found, but each ion behaved like a molecule in any process related to vapor pressure.

At present, we agree with Arrhenius, that water breaks up molecules of acids (and how it does this will be handled later in the book), but with hydroxides and salts, it merely causes the ions that are already there to dissociate. The "degree of ionization," so important when Arrhenius formulated his scheme, turns out to be not quite what it seemed, except for weak acids.

The distinction between acids and the other two classes is now often made clearer by the use of different words. Acids are said to *ionize* in water, but hydroxides and salts *dissociate*. This distinction is still not universally made, but I shall try to stick to it in this book.

# 8

## The Solubility Product: Teaching

## the Elementary Mathematics of Equilibrium*

If you've never seen it done, try this simple experiment. Make a saturated solution of common salt by shaking up a lot of it with water. When the excess solid has settled, pour off the clear solution. Now add a little concentrated hydrochloric acid to this, and mix it. A heavy, white precipitate will form. A somewhat more dramatic way to do the same thing (it makes a good magic show) is to make hydrogen chloride gas, and pipe it in on top of the saturated salt solution: the same precipitate will form, as a snowstorm of white crystals. See Figure 8-1 on following page.

What is the precipitate? If you look at solubility tables, you find that there is no "insoluble" substance that could be formed from the NaCl-HCl combination.

Now vary the experiment a little. To a new lot of saturated salt solution, add a little concentrated sodium hydroxide solution (preferably 10M or stronger) and mix. Again a precipitate appears, even though both NaCl and NaOH are very soluble in water.

The principle illustrated in both these demonstrations is called

---

*Portions of the material included in this chapter first appeared in the journal *Chemistry* and are reprinted here by permission of the American Chemical Society.

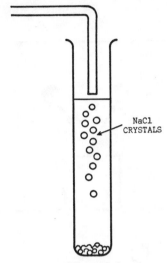

**Figure 8-1**

the *common ion effect*. If you start with a saturated solution of an ionic compound, and add something that raises the concentration of *one* of its ions, some of the solute will crystallize out. In both of these experiments, the precipitate was simply salt, NaCl.

The common ion effect is one of a wide range of phenomena involving equilibria among ions. The basic methods of treating these are very similar to the ones we have used with liquid-vapor systems: they involve Le Chatelier's Principle, and consideration of reaction rates and entropy. And as we did when we were working with freezing and boiling point effects, we'll take up here the three questions: What happens?, Why?, and How much?

We begin by looking at the salt solution as it was being prepared. When we first put the solid salt into water, sodium and chloride ions started escaping from the salt crystals. See Figure 8-2. At first they diffused widely through the container, but soon some of them began to return to the solid state, reattaching themselves to the surface of the crystals in pairs. The return rate was slow to begin with, because there weren't many ions in solution; but as time (and stirring) went on, the concentration of ions increased to the point at which the rate of return exactly equaled the rate of escape. See Figure 8-3. The solution was now saturated: the total amount of dissolved salt remained constant.

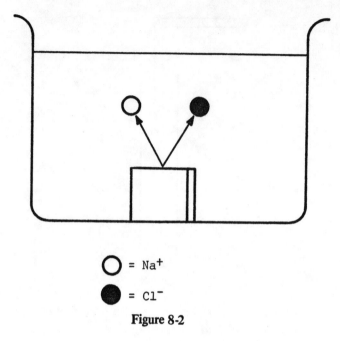

= Na$^+$

= Cl$^-$

**Figure 8-2**

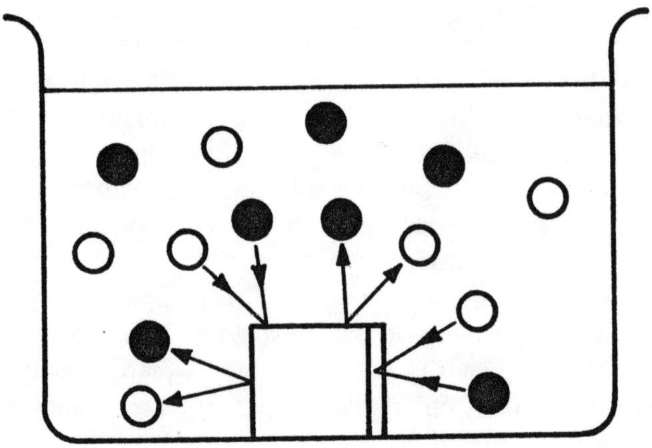

**Figure 8-3**

This saturation was a dynamic condition. The *number* of ions in solution didn't change, but their identity did: individual ions were engaged in furious activity, escaping from and returning to the crystals. The underlying cause of the *escape* process (as with the evaporation of a liquid) is the tendency of any system to become disorderly—for its entropy to increase. The cause of the *return* of ions to the crystal is of course the tendency of opposite charges to get as close together as possible—that is, to assume a state of lowest energy. As we said earlier, the balance between energy and entropy determines the direction of any dynamic process, and at equilibrium their changes are equal.

*Le Chatelier's Principle*

The first and perhaps least informative way to account for what happens when HCl is added to this system is to use Le Chatelier's Principle. When we added strong hydrochloric acid to the sodium chloride solution, the boost in chloride ion concentration represented a stress. The system then had to react to decrease this stress—to remove some of the added chloride ions. This could happen in the context of our equilibrium only if the chlorides combined with sodium ions, and dropped out of the solution as salt crystals. Essentially, the same thing happened when concentrated sodium hydroxide was added: some of the excess sodium ions (the stress) were removed, again by the formation of salt crystals.

Le Chatelier's Principle turns out to be very useful, again and again, for predicting *what* will happen, but it tells us nothing about *why*. So let's look at the experiments from the point of view of collisions. We'll concentrate on the HCl addition, since the principle is the same in both experiments.

*Collisions*

I've already said that in the saturated solution, sodium ions and chloride ions are arriving at the surface of the crystals at the same rate as other sodium and chloride ions are escaping. They must, on the average, collide with the salt surface in pairs, otherwise disruptive electrical charges would build up. The rate at which each kind of ion arrives clearly depends on its concentration in the solution. When we add extra chloride ions, by pouring in HCl, this increases the probability that a chloride ion will arrive at the surface at the same time as a sodium ion. See Figure 8-4. All this doesn't change the rate of *escape* of ions from the crystal surface, so the crystal grows.

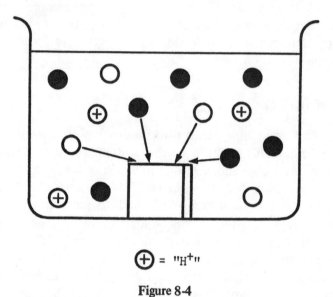

$\oplus$ = "$H^+$"

Figure 8-4

This analysis of the experiment leads directly to the next way of handling it. So far we've talked in qualitative terms only. With Le Chatelier, we found what would happen. The collision theory told us why it would happen. Now we want to know how much it will happen.

*Constants*

We've just seen that an increase in the concentration of one of the ions in the saturated solution increases the likelihood of collision between a sodium ion and a chloride ion. Let's look at this likelihood factor quantitatively. Start with a visible analogy: imagine a room the size of a basketball court. In it are two men and two women, all of them blindfolded. See Figure 8-5. The

Figure 8-5

game is for a man and a woman to find each other by collision. When they do, they may retire from the floor. If two people of the same sex bump into each other, they separate disgustedly. If the game is repeated again and again, the average time for the first effective collision can be determined.

Figure 8-6

Now vary the game by adding two more men, to make four, but still with two women. See Figure 8-6. Because each woman has twice as many possible men to bump into, the time for the first effective collision should be cut in half. If there are six men to two women, the time will be one-third what it was originally. Clearly, the rate of effective collisions varies directly with the number of men.

Now reverse the procedure: start with two-and-two as before, triple the collision rate by tripling the number of men only, and then double the number of women. The effective collision rate will be six times what it was in the beginning. The law governing their collisions is simple enough: the rate is proportional to the *product* of the number of men and the number of women on the floor; that is:

$$\text{Rate} = k(N_m)(N_w).$$

The application to ionic collisions is obvious. Collisions between ions of like charge are ineffective, but when oppositely charged ions meet, crystallization can occur. Therefore, the crystallization rate is proportional to the product of the concentrations of the sodium and the chloride ions. In a saturated solution, this rate is equal to the rate of escape of ions from the solid (which, as I said, is constant, driven by the tendency to disorder). Therefore, the ion-product in a saturated solution is a constant, called the solubility product, $K_{sp}$, for the salt.

Now is the time to clarify a matter of terminology that I mentioned in an earlier chapter. Up to now I have spoken of *molal* solutions, and I said that these differ slightly from mo*lar* solutions. In this chapter we have been dealing with interactions of ions with one another, in contrast to the situations dealt with earlier where molecules or ions were interacting with their *solvent*. The difference is that we are interested now in the total volume—the size of the basketball court. If you add a mole of sugar to 1000 grams of water, as we did in working with the freezing point, the total volume will be considerably more than the 1 liter that the water had originally occupied. But our basketball court analogy is good only if the court doesn't change in size. A 1-mo*lar* solution, then, is one that contains 1 mole of the solute in 1 liter of total solution. Making it is easy enough, we simply dissolve 1 mole of the solute in "some" water—less than a liter—and then add water until the final volume is exactly a liter.

For very dilute solutions, the distinction between molar and molal is usually unimportant. For instance, a solution of hydrochloric acid that is 0.1000 molal will be about 0.0997 molar at room temperature. And a 0.1000 molal ammonia solution is about 0.1001 molar. For very precise work these distinctions are important, and even the symbols for the two are different—small m for molal, and capital M for molar. They will not be important for anything we do here; I shall simply be careful to use the proper word and symbol.

I'll introduce here another universally used symbol, the square bracket—e.g., $[Na^+]$. This means "molar concentration of sodium ions." In a solution of sodium chloride, the ion-product is $[Na^+] \cdot [Cl^-]$, and in a saturated solution, this product is a constant, the solubility product.

At this point I should note that I've oversimplified in two ways in describing the situation in a saturated solution. First, I have not clearly distinguished between collisions of ions in solution and collisions at a crystal surface. This distinction could get us into the tricky area of supersaturation, which is better left for treatment elsewhere. And, second, I have not taken the surface area of the crystal into account. It turns out that this latter item doesn't matter: the rate of escape of ions from a surface obviously will vary with its size—but so will the rate of return. Consequently, this factor has no effect on the equilibrium.

## Calculations

A saturated solution of NaCl in pure water contains about 5 moles of salt per liter. If we assume that the salt in solution is completely dissociated, there will be 5 moles of each ion there, and the $K_{sp}$ is 5 x 5 or 25. If we could double the chloride concentration without changing the volume (we can almost do this by piping in HCl gas, which doesn't change the volume very much), we could double the rate of collisions, and a lot of salt would crystallize out. Of course, in this process, a lot of the excess chloride ion would drop out too. However, let's keep on adding HCl gas until the *final* chloride concentration is 10M; then the sodium ion must drop to 2.5M (since 10x2.5 = 25) to keep the rate of crystallization equal to the rate of escape of ions from the crystals. This means that 2.5 moles, or half the NaCl originally present, will have dropped out.

If the HCl could be boosted to 25M (it can't), only 1 mole of sodium ions would remain in solution and four-fifths of the original NaCl would have crystallized.

Here another note on oversimplification. I have assumed that the solubility product rule applies exactly in all cases, and also that it is possible to add a lot of extra ions to a solution without changing its volume. Neither of these assumptions is correct, though in the case of salts with small values of $K_{sp}$ they can be closely approximated. I'll go on using sodium chloride in this discussion for convenience, and I'll take up volume changes in a minute. But the principles discussed here do apply, with only very slight modifications, to many practical situations.

### Stress vs Collisions

Notice that the Le Chatelier explanation and the solubility product explanation for our original demonstration say slightly different things. Le Chatelier says that because the added chloride creates a stress, the system reacts in such a way as to decrease the number of these added *chloride* ions and that is why salt precipitates. The $K_{sp}$ explanation says that, because extra chloride ion has been added, the system will react to decrease the *sodium* ion by precipitating salt. The explanations are definitely different, even though the results are identical.

### Reversal

In adding extra ions to a saturated solution, we also necessarily increase the volume. This volume change has been neglected in the cases we've dealt with so far, but let's try another experiment. Take 1 liter of the original salt solution containing 5 moles each of sodium and chloride ions. To this add a total of 1 liter of concentrated (assume 12.5M) HCl solution—and this time be realistic about the volume.

When the first of the acid solution is added, the extra chloride ions make salt precipitate. But after adding *all* the acid, this salt should dissolve up again. At this point we'll have 2 liters of solution, containing all the ions originally present in both solutions: 5 moles of the original sodium ions, and a total of 17.5 moles of chloride ions. Thus their concentrations *per liter* are: $[Na^+] = 2.5M$, and $[Cl^-] = 9M$ (approx.). The product of these is about 23, which is less than the solubility product, so there should be no precipitate, and the crystals that formed first, when a *little* HCl was added, should redissolve. Further addition of concentrated HCl will dilute the sodium ions more than it will increase the chloride ion concentration; therefore, no matter how much more HCl is added, no salt should precipitate.

Le Chatelier's explanation handles this situation equally well. At first we stressed the system by increasing the chloride ion concentration. This stress was relieved by the formation of the crystals. Later, a new stress entered the picture: we *decreased* the *sodium* ion concentration to an important degree. The system obediently countered this new stress by letting the NaCl redissolve.

## Electrolytes Containing More than Two Ions

Although we have dealt in detail here with only a single case, and a simple one, the basic ideas apply to a wide range of compounds. Only minor changes need be made to handle other cases. It turns out that the product that remains constant in the case of a salt like $Ag_2CO_3$ is $[Ag^+]^2 \cdot [CO_3^=]$; and for $Ca(OH)_2$, the $K_{sp}$ is $[Ca^{++}] \cdot [OH^-]^2$. In a saturated solution of calcium hydroxide, its concentration is a little over 0.02M. This means that $[Ca^{++}]$ is 0.02, and $[OH^-]$ is 0.04. Then the $K_{sp}$ = $[Ca^{++}] \cdot [OH^-]^2$ = 0.000032, or a little over. (Most textbook tables, by the way, give an incorrect value for this constant—I don't know why.)

The determination of the value of $K_{sp}$ for calcium hydroxide is a challenging lab exercise for students who have learned acid-base titration. Simply prepare, and carefully standardize, solutions of sodium hydroxide that are roughly 0.1N, 0.05N, and 0.025N (i.e., 1/10, 1/20, and 1/40). After standardizing, shake up an excess of powdered calcium hydroxide in each solution, let it settle (or filter it) and titrate the resulting saturated solution. Also prepare a saturated solution of calcium hydroxide in pure water and titrate this. From the results of these titrations, it is possible to calculate, in each case, the final $[Ca^{++}]$ and $[OH^-]$, and from these find the value of $K_{sp}$.

Two notes on this exercise: First, as stated above, the value of the constant is usually given as $1.3 \times 10^{-6}$, which is definitely false. Second, with sufficiently precise work (it has happened a few times with some of my students), the values can be good enough to show the Debye-Hückel effect: the solubility product *rises* very slightly with increasing total ion concentration of the solution.

## Some Puzzles

Silver acetate has a solubility product of about $2.5 \times 10^{-3}$, low enough to follow the rules quite well, yet high enough so crystallization can be easily observed. Put a little saturated silver acetate solution in a test tube, and add a pinch of silver nitrate crystals. *Immediately* swirl it and watch closely. The added crystals will quickly dissolve; then a new type of crystal, finer and flakier, will appear. Why?

To make silver acetate, which I find difficult to buy, I usually start with silver nitrate, which is easily available. Why doesn't it work to take a liter of 1M silver nitrate solution and add a mole of pure acetic acid to it (a saturated solution of silver acetate is only about 0.05M); whereas, it works well to add a mole of pure sodium acetate?

In this same preparation of silver acetate, I usually add, not 1 mole of sodium acetate, but two. Why "waste" sodium acetate?

## Exercises

Calculations with solubility products show in some cases the kind of accuracy a scientist is satisfied with. In other cases, they indicate ingenuity of method sometimes needed in attacking a problem. We'll consider one simple problem here, and then a couple that have greater interest.

*Exercise 1. How many grams of calcium sulfate should dissolve in 400 ml of water? Its $K_{sp}$ is $2.4 \times 10^{-5}$.*

Solution. In a solution containing nothing but $CaSO_4$, the concentration of each of its ions must be the same; call it x. We are given that

$$[Ca^{++}] \cdot [SO_4^{=}] = 2.4 \times 10^{-5}$$

so in this case

$$x^2 = 2.4 \times 10^{-5}$$

and

$$x = 4.9 \times 10^{-3}$$

That means that this number of moles of *each* ion, or $4.9 \times 10^{-3}$ moles of $CaSO_4$, will dissolve in a liter. One mole = 136 g. Therefore $(4.9 \times 10^{-3})(136)$ g dissolves in a liter, and $\frac{400}{1000} \times (4.9 \times 10^{-3})(136)$ g = 0.26g in 400 ml. Notice that throughout this calculation I've stuck to two significant figures. This is normally the maximum accuracy allowable with solubility products.

*Exercise 2. If the $K_{sp}$ of magnesium oxalate is 8.6 x $10^{-5}$, how many moles of this salt should dissolve in a liter of 0.3M magnesium chloride solution?*

Here we have a slight complication: the number of moles of *oxalate* in the final solution will depend only on the amount of magnesium oxalate that dissolves: but for the *magnesium*, we must think about the amount that dissolves *plus* the much larger amount already in solution. What makes the problem easy is those words "much larger." We know, from the small size of the $K_{sp}$, that only a little magnesium oxalate will dissolve. As it does so, it will of course increase the amount of $Mg^{++}$ ion—but *not enough to matter*. So in our calculation we'll simply forget about (the mathematician says "neglect") this added magnesium, and assume that after magnesium oxalate has dissolved, the $[Mg^{++}]$ will still be 0.3M.

Now the calculation is simple—could even be done in your head.

$$[Mg^{++}] \cdot [C_2O_4^{=}] = 8.6 \times 10^{-5}$$

or $\qquad$ $0.3x \qquad = 8.6 \times 10^{-5}$

$\qquad\qquad x \quad = 2.9 \times 10^{-4}$ $\qquad$ and about 0.0003

moles of magnesium oxalate will have dissolved.

The size of this figure shows that we were justified in neglecting the added magnesium in doing the calculation. If we *had* considered it, there would be a change of the order of 1% in the value of $[Mg^{++}]$, and this is far below the accuracy of the $K_{sp}$ figure itself. I'll go into some detail on "scientific cheating" of this sort when we study buffers, where it is used extensively.

*Exercise 3. How much silver acetate ($K_{sp}$ = 2.5x$10^{-3}$) should precipitate from a liter of saturated solution if half a mole of solid sodium acetate is added?*

There are two ways of attacking this problem; one is easy, the other hard. For either way, we must first find out how many moles of silver acetate are present to begin with. We could then (the hard way) let x equal the amount precipitated, and get rather unpleasantly bogged down in a quadratic equation (which would have to be solved to four significant figures to get a two-significant-figure result). Try it if you enjoy calculations.

But rather than doing this, we'll use an indirect approach. We'll suppose that we start from scratch, make up a 0.5M sodium acetate solution *first,* and then dissolve solid silver acetate in this. The resulting final concentration of silver acetate will be the same as it would have been if we'd added the solids in the other order, and the method of calculation will be exactly the same as the easy method of Exercise 2. Then, having found the amount in solution, simple subtraction gives us the amount that would precipitate if we did the experiment as given.

Initially, in the saturated solution (letting x = moles of silver acetate dissolved): $x^2 = 2.5 \times 10^{-3}$.

or        $x = 5 \times 10^{-2}$. Thus 0.05 moles of silver acetate is present in the original solution.

Now, if silver acetate were dissolved in 0.5M sodium acetate solution (letting y = moles of silver acetate that dissolves in this case, and neglecting its effect on the acetate ion concentration) (see Exercise 2):        $0.5y = 2.5 \times 10^{-3}$?

Then        $y = 5 \times 10^{-3}$.

But if only about 0.005 moles of silver acetate can be in solution when the acetate concentration is 0.5M, the rest of the silver acetate originally present must have precipitated; that is, $0.05 - 0.005$, or 0.045 moles. Roughly 90% is forced out by the acetate ion of the sodium acetate. I say "roughly," but a careful check will show that the error of our method is of the order of only 1%, which is less than the uncertainty of the $K_{sp}$ value.

This exercise, by the way, contains the answer to the third "puzzle" on page 112.

### Conclusion

Solubility product considerations are of both theoretical and practical importance. They can tell us which precipitates will form when solutions are combined, and are therefore crucial in many chemical analyses (see especially Chapter 13). They make some commercial processes possible (see Chapter 12). Soil fertility is influenced by acidity, which in turn is related to the solubility of the phosphates. And the mineralogy of limestone, from the stalactites of the Luray Caverns to the hard water that clogs an electric iron, is connected with the solubilities of the carbonates (see Chapter 9).

**9**

## Acids, Bases, and Equilibrium:
## How a Theory Grows

As science advances, each theory ordinarily goes through three important stages. When first proposed it is often doubted, sometimes even laughed at. In the face of this, its supporters test it against all kinds of experimental facts. If it survives this stage, the new "Theory" begins to be written with a capital letter, and almost takes on the status of a fact. Then small inaccuracies or exceptions begin to pop up. These require modification of the theory; its use gets more complex. It may still be retained, for convenience; or some or all of it may be discarded, to be replaced with something simpler, more comprehensive, or more accurate. Most students are familiar with these stages for theories of the universe, from Ptolemy to Copernicus, and on to Kepler, Newton, and Einstein. Here we'll follow some of the stages for theories of acids and bases.

In Chapter 7, we began to look at what happens to an acid in water solution. We carried our treatment through to Arrhenius' theory that acids ionize as a result of the action of water on their molecules. This theory has held up well, but with clearer understanding of this acid-water interaction, more general systems have been developed which include the Arrhenius one as a special case.

The need for more general theories arose in connection with a number of observations. For one thing, hydrogen ions in water seemed to be much too big—they move too slowly really to be single ions with a mass of 1 amu. Then some salts, like sodium carbonate, are quite strongly basic in water solution, while others, like ammonium or aluminum chloride, are acidic. Arrhenius' theory could be modified fairly successfully to handle some of these, but it was less successful with others, and began to get complicated. And the Arrhenius system didn't work at all when we were dealing with solutions in liquid ammonia, concentrated acetic acid, and other non-aqueous solvents.

The *Brønsted-Lowry* acid-base theory handles all of these easily and simply, and we'll go into it in some detail. We'll merely touch on the *Lewis* theory, which has the broadest application, but tends to be complicated in its applications.

### Hydronium Ions

To begin with, it turned out that Arrhenius' hydrogen ions in water aren't quite what he supposed them to be. Every charged particle in water will collect a cluster of water molecules around it, since water molecules have oppositely charged "ends." But, more than this, the hydrogen ion has a special relationship to the water molecule.

A hydrogen ion is in fact simply a single proton—a tiny, very highly charged, very dense blob of matter. And it is identical with the two protons already attached by covalent bonds to the water molecule. If we imagine a proton, lost in space, happening on to a water molecule, this will look just like home to it. Two diagrams here show the situation. One is a schematic one, showing electrons; the other is drawn to a more accurate scale (but still crudely). See Figure 9-1. The free proton will snuggle up to the oxygen atom, covering itself cozily with one of the electron pairs so readily available, and from then on it cannot be distinguished from the two protons that previously had exclusive occupancy. The ion that results from this successful squatting is called "hydronium";* it has an unbalanced charge of +1, but is otherwise very stable.

---

*The name *hydronium* was coined for the similarity of this ion to *ammonium*, the two being formed in the same way.

$$\overset{+}{H} + \overset{\cdots}{\underset{H}{:O:}}H \rightarrow H\overset{\cdots}{\underset{H}{:O:}}\overset{+}{H}$$

OR

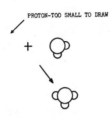

PROTON—TOO SMALL TO DRAW

**Figure 9-1**

## Ionization of Water

Of course wandering protons are rare, but even attached protons sometimes have a wandering eye. In pure water itself, there is a small proportion of positively charged hydronium ions. They are formed when a pair of water molecules happen to bump into each other in exactly the right positions, and at the right speed, to exchange a proton. See Figure 9-2. In the collision, a hydroxide ion is produced from the water molecule that has lost the proton.

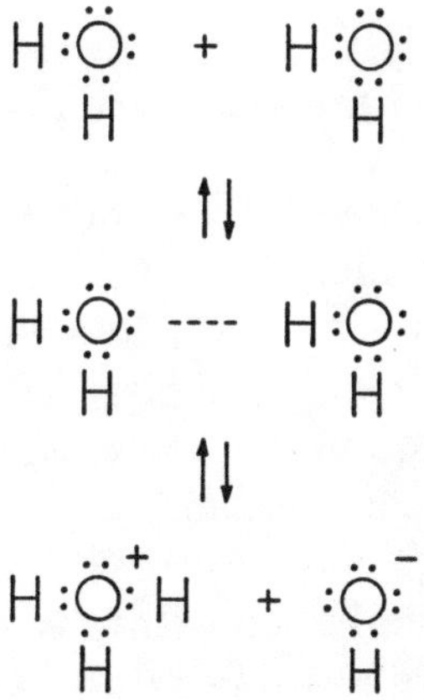

$$H\overset{\cdots}{\underset{H}{:O:}} + H\overset{\cdots}{\underset{H}{:O:}}$$

$$\updownarrow$$

$$H\overset{\cdots}{\underset{H}{:O:}} ---- H\overset{\cdots}{\underset{H}{:O:}}$$

$$\updownarrow$$

$$H\overset{\cdots}{\underset{H}{:O:}}\overset{+}{H} + \overset{\cdots}{\underset{H}{:O:}}^{-}$$

**Figure 9-2**

In water, the life of these ions is short: each will soon bump into another of opposite charge, and re-exchange a proton to make a new pair of water molecules. So here we are again, with a condition of dynamic equilibrium. These processes are constantly going on, and at any given time, in a liter of pure water, at room temperature, there are almost exactly $10^{-7}$ moles of hydronium ions and the same number of hydroxide ions. The amounts may seem tiny, but they are exceedingly important.

## Acids in Water

When hydrogen chloride is introduced to water, the same proton exchange happens, but on a massive scale: almost every one of the HCl molecules loses its respective proton, and the result is a solution containing almost nothing but hydronium ions and chloride ions:

$$HCl + H_2O \longrightarrow H_3O^+ + Cl^-.$$

In a solution of a weak acid, such as acetic, we have a situation more like that in pure water: a few acetic acid molecules lose protons to water and quickly regain them. We saw in Chapter 7 that when equilibrium is attained, about 0.5% of the acetic acid originally present (with a 1M solution) has lost its protons.

As far as *calculations* go, the distinction between hydrogen ions and hydronium ions is unimportant. One mole of protons will yield exactly 1 mole of hydronium ions. Many scientific writers economize by writing $[H^+]$ instead of $[H_3O^+]$.

## The Ionization Constant for Water, and the pH Scale

The fact that hydronium ions are present in water itself, as well as in solutions of "Arrhenius" acids, calls for a simple way of handling the ions. Our work with solubility products furnishes a strategy. We found there (Chapter 8) that when oppositely charged ions were in equilibrium with crystals, the concentrations of the dissolved ions could be multiplied to give a constant, the solubility product, $K_{sp}$.

In water there is a similar constant, as a result of a similar kind of equilibrium. We have a steady rate of *formation* of hydronium and hydroxide ions directly from water, practically unaffected by anything else that is going on. But hydronium and hydroxide ions will recollide at a rate that is proportional to the product of their concentrations (just as positive and negative ions do in a salt solution). Therefore, we should expect to find—and do—that $[H_3O^+][OH^-]$ is constant. The value of the constant, $K_w$, can be found from a figure that I've just given: in pure water the concentration of each of the two ions is $1.0 \times 10^{-7}$, and therefore $K_w$ is $1.0 \times 10^{-14}$. This value will remain constant in any water solution, no matter what else is there (acids, salts, etc.), as long as there is approximately 1 kilogram of water in a liter of the solution; i.e., in all dilute solutions.

And, since this product remains constant in any water solution, we have a convenient way to deal with the concentrations, not only of acids, but also of bases. In a 0.1M solution of hydrochloric acid, $[H_3O^+]$ is about $10^{-1}$ (notice that the hydronium contributed by water itself is not significant relative to this figure). If the product, $K_w$, is to remain constant, $[OH^-]$ must now be only $10^{-13}$. (For a pictorial analysis of the collisions that yield this result, see Chapter 11 on Buffers.) Likewise, in 0.01 M sodium hydroxide solution, the $[OH^-]$ is $10^{-2}$, so $[H_3O^+]$ must be $10^{-12}$.

Small changes in the concentrations of these two ions can have powerful effects on chemical reactions, especially those that occur in living things. Around 1920 biologists got tired of writing down all the negative exponents, and started using a term called the pH. The symbol came from "potential of hydrogen" and was first used by S.P.L. Sφrensen in 1909. For practical purposes, especially rough ones, you look at the hydronium ion concentration, expressed as a power of 10—and then leave out the 10 and the negative sign that is usually there. Thus if $[H_3O^+]$ is $10^{-9}$ the pH is 9, and if $[H_3O^+]$ is $10^{-12}$ (as it is in 0.01M NaOH) the pH is 12. Notice that the higher the acidity the lower the pH, and that in neutral solution the pH is 7.

Formally (and importantly for precise work) the pH is defined as the negative logarithm of the hydronium ion concentration. This takes care of cases where the power of 10 is not an integer. Thus if $[H_3O^+]$ is $2 \times 10^{-4}$, the pH is $-(\log 2 + \log 10^{-4}) = -(0.30 - 4.0) = 3.7$.

## The Brφnstead-Lowry System

With that background, we're ready to handle the first of the newer acid-base theories, proposed by Jφhannes Bronsted and

Thomas Lowry in 1923. Their theory defines an acid as a molecule *or ion* that can lose a proton–a "proton donor"; and a base is a molecule or ion that can gain a proton–a proton acceptor.

Since all reactions are to some extent reversible, every acid-base reaction yields a new acid-base pair. Consider a simple case that would not have been called an acid-base reaction in Arrhenius' scheme, the dissolving of HCl gas in water. The older theory would have the molecules simply ionizing, to give $H^+$ and $Cl^-$ ions. Modern theory has the HCl give a proton to the water, *which therefore acts as a base:*

$$HCl \quad + \quad H_2O \rightleftharpoons H_3O^+ \quad + \quad Cl^-$$

Acid (strong)    Base (weak)    Acid (Strong)    Base (weak)

Since the hydronium ion can return its proton to the chloride ion (it does so easily only in concentrated hydrochloric acid, which always has a strong smell due to HCl molecules escaping), these last two ions are also respectively acid and base. When an acid loses a proton to form a base, the "pair" are called "conjugates." In the immediately preceding reaction, HCl and

| Some conjugate acid-base pairs | |
|---|---|
| Acid | Base |
| $H_2O$ | $OH^-$ |
| $H_3O^+$ | $H_2O$ |
| $OH^-$ | $O^=$ |
| $NH_4^+$ | $NH_3$ |
| $H_2SO_4$ | $HSO_4^-$ |
| $HSO_4^-$ | $SO_4^=$ |
| $Cu(H_2O)_4^{++}$ | $Cu(H_2O)_3OH^+$ |

$Cl^-$ are conjugate acid and base, and $H_2O$ and $H_3O^+$ are conjugate base and acid. The table here shows some other typical acid-base pairs. Notice that acids and bases can be neutral molecules, or positively or negatively charged ions. *Every* negatively charged ion can act as a base, since it can accept a proton.

Arrhenius would have considered the neutralization of hydrochloric acid with sodium hydroxide solution as the only true type of acid-base reaction; the Brønsted treatment of this is not very different from his:

$$H_3O^+ \quad + \quad OH^- \rightleftharpoons H_2O \quad + \quad H_2O.$$

Acid (s)    Base (s)    Base (w)    Acid (w)

In solution, both of the compounds are in the form of ions (HCl has "ionized" and NaOH has "dissociated"), and the only reaction that occurs is the one shown. Two things are significant about this reaction. First, when both acid and base are strong, the reaction proceeds very far to the right, even though it is reversible. Second, we see here one kind of molecule, water, acting as both acid and base. Water is only one example of the many molecules and ions that are able to act in either capacity, depending on conditions. Notice that $HSO_4{}^-$ appears in both columns of the preceding table.

## Polyprotic Acids

This "amphiprotic" (i.e. either acid or base) nature of certain materials is clearly shown by an experiment. Suppose you were to take a 1M solution of phosphoric acid and add 1M sodium hydroxide solution to it. Elementary chemistry will tell you that 3 volumes of the basic solution will be needed for every volume of the acidic one if you want to make sodium phosphate:

$$H_3PO_4 + 3NaOH \longrightarrow Na_3PO_4 + 3H_2O.$$

Try it. Put exactly 10 ml of 1.0M phosphoric acid solution in a small flask, add a few drops of methyl orange indicator (this is normally available—we'll see in a moment why this particular one was chosen), and titrate the acid solution with 1.0M sodium hydroxide. Use a burette or graduated pipette.

The acid solution starts with a deep orange color. The color changes to yellow when 10 ml—not 30!—of the base has been added. The change is a very sharp one: one drop of the base at the "endpoint" will make it happen. That is, there is a clear "neutralization" point when only a third of the predicted quantity of sodium hydroxide has been added. If the solution were evaporated at this stage, you would find a single, pure crystalline salt.

But continue with the titration. Add to the same solution a few drops of either phenolphthalein or thymol blue indicator (either will work; I prefer the latter) and again add sodium hydroxide solution. There will be no further change in color until a second 10 ml of this has been added. Then, again, very suddenly, the color will change from yellow to bright orange (if phenolphthalein was used) or green (with thymol blue). Again there has been a definite "neutralization," this time with two-thirds of the predicted quantity of base. Again, if the solution were evaporated to dryness, a single kind of crystal would be found, different from that obtained at the first stage.

For the third stage of neutralization, when finally 30 ml of sodium hydroxide has been added, there is no practical indicator.

It would be finally at this point that evaporation of the solution would yield sodium phosphate, $Na_3PO_4$.

At what point in the process could you say that the acid had actually been "neutralized"? The question has no real answer. The first indicator used, methyl orange, changes color at about pH 4; that is, when the solution is still quite acidic. Phenolphthalein and thymol blue change at about pH 10; that is, when the solution is quite basic. And the final sodium phosphate solution is so strongly basic that trisodium phosphate is sold as TSP, a cleaner and paint-stripper.

Consider the separate stages of the process from the point of view of the Brønsted-Lowry theory. Throughout the reaction the sodium ions are mere "spectators," taking no part unless the solution is evaporated. But upon evaporation, three distinct sodium salts will be obtained: $NaH_3PO_4$, $NaH_2PO_4$, and $Na_3PO_4$.

(a)   $\underset{\text{acid}}{H_3PO_4}$   $+$   $\underset{\text{base}}{OH^-}$   $\longrightarrow$   $\underset{\text{base}}{H_2PO_4^-}$   $+$   $\underset{\text{acid}}{H_2O}$

(b)   $\underset{\text{acid}}{H_2PO_4^-}$   $+$   $\underset{\text{base}}{OH^-}$   $\longrightarrow$   $\underset{\text{base}}{HPO_4^=}$   $+$   $\underset{\text{acid}}{H_2O}$

(c)   $\underset{\text{acid}}{HPO_4^=}$   $+$   $\underset{\text{base}}{OH^-}$   $\longrightarrow$   $\underset{\text{base}}{PO_4^{\equiv}}$   $+$   $\underset{\text{acid}}{H_2O}$

Notice that in both (b) and (c) the "base" that resulted from the preceding reaction is now acting as an acid. This is precisely what the Brønsted theory is about. The question of whether an amphiprotic molecule or ion is an acid or base depends entirely on the conditions it is placed in. The two hydrogen-phosphate ions, $H_2PO_4^-$ and $HPO_4^=$, can so easily swing either way that they are among the most useful tools of the biochemist who wants to maintain the pH of a solution in the neighborhood of exact neutrality. "Buffer solutions" for this purpose will be discussed in Chapter 11.

The acid-or-base nature of an amphiprotic substance can be demonstrated effectively with one of the acid phosphates we've just been discussing—sodium dihydrogen phosphate, $NaH_2PO_4$ (sometimes labeled "sodium phosphate, monobasic"). This material will visibly "neutralize" either a strong acid or a strong base.

For indicator, we'll use thymol blue again (see the experiment at the end of this chapter for the preparation of this indicator, and its pH ranges). In strong acids the indicator is red. It is yellow when the solution is nearly neutral, but still acidic; and turns blue in basic solutions.

Take a test tube half full of 0.1N hydrochloric acid, and add a drop or two of thymol blue solution. The liquid will turn bright red. Now add half a gram or so of solid $NaH_2PO_4$ and swirl the tube. As the solid dissolves, the color will slowly change from red to yellow, indicating that the acid has been "neutralized" by the salt. Strictly speaking, of course, it has been made less acidic.

Next, start with 0.1N sodium hydroxide solution in the same way, with the same indicator. The solution will be bright blue. Again add solid $NaH_2PO_4$ and swirl the tube. Again the indicator color will change to yellow, showing that the *base* in turn has been "neutralized." It is true that the final solutions in the two cases don't have the same pH. But both end up in the pH region between 3 and 8, a little on the acid side of precise neutrality.

In both of these cases, the same ion, $H_2PO_4^-$, has acted. With hydrochloric acid, its reaction was:

$H_3O^+ + H_2PO_4^- \longrightarrow H_3PO_4 + H_2O$, and the resulting phosphoric acid doesn't ionize enough under these conditions to affect the indicator.

With sodium hydroxide, the reaction was:

$OH^- + H_2PO_4^- \longrightarrow HPO_4^= + H_2O$, and the $HPO_4^=$ ion is nearly neutral.

Another demonstration of the same type can be done with baking soda (sodium dihydrogen carbonate, sodium bicarbonate), although the demonstration goes best with an indicator that is less often on hand, azo violet. This indicator will change color between about pH 11 and 13.

When this indicator is added to 0.1N sodium hydroxide solution, and solid baking soda is added, there is a clear color change, from purple to yellow, as the pH drops. The reaction here is:

$OH^- + HCO_3^- \longrightarrow CO_3^= + H_2O$. The special indicator is needed because the carbonate ion is itself rather a strong base, as we'll see in the next section.

To neutralize an *acid* with baking soda, 0.1 N hydrochloric acid with thymol blue indicator can again be used. When the solid salt is added there is violent fizzing, and the color changes from red to yellow. Here the reaction is:

$H_3O^+ + HCO_3^- \longrightarrow H_2CO_3 + H_2O$. The carbonic acid immediately decomposes, creating bubbles of carbon dioxide gas.

*Hydrolysis: Basic Solutions*

The amphiprotic nature of water itself is responsible for the fact that many salts give solutions that are not neutral. The phosphates just discussed show this fact very clearly, but it can be handled more simply if we back-track a little and deal with the monoprotic acetic acid. Sodium acetate solutions are always slightly basic; that is, the $OH^-$ ion concentration is always slightly higher than the $H_3O^+$ ion, or the pH is always above 7. This is caused by the reaction of the acetate ion with water.

Like all negative ions, the acetate ion is a base. Its conjugate acid is acetic acid. Since this is a weak acid, the acetate ion is a moderately strong base (and the terms "strong" and "weak" must be considered very rough ones from here on).

Therefore there is a slight tendency for the following *hydrolysis* reaction to occur:

$$C_2H_3O_2^- \quad + \quad H_2O \rightleftharpoons HC_2H_3O_2 \quad + \quad OH^-$$

Base(s)          Acid (w)     Acid (w)        Base(s).

Since $OH^-$ is the strongest base that can exist in water solution, this reaction does not go very far to the right, but enough so to make the pH of a 0.1M sodium acetate solution a little below 9.

Any negative ion can "steal" a proton from water in this way, and hydrolysis of this type is important for salts of all acids, except the very strong ones—hydrochloric, nitric, sulfuric, and a couple of others. Conjugate bases of these very strong acids are exceedingly weak bases, which do not hydrolyze to any important extent.

Acids that can lose two or three protons may give salts that hydrolyze very strongly. Phosphoric acid, $H_3PO_4$, will ionize in three stages, becoming more reluctant at each stage to give up the proton:

$$H_3PO_4 \longrightarrow proton + H_2PO_4^- \longrightarrow proton + HPO_4^=$$

$\longrightarrow proton + PO_4^\equiv$. Consequently, the $PO_4^\equiv$ ion is a base comparable to $OH^-$ in strength, and a solution of trisodium phosphate (commonly sold as TSP) can be used to remove paint, or scrub up grease, with the high $OH^-$ concentration in solution:

$$PO_4^\equiv + H_2O \rightleftharpoons HPO_4^= + OH^-.$$ A 1M solution of sodium phosphate is about 10% hydrolyzed; that is, it has as high an $OH^-$ concentration as 0.1M sodium hydroxide; and sodium sulfide, $Na_2S$, can be more than 90% hydrolyzed.

## Hydrolysis: Ammonium Salts

Hydrolysis of salts to give acidic solutions is a little more complicated. We'll begin with those of ammonia, which is rather exceptional in itself. Ammonia is a weak base, in the Brønsted sense, ionizing slightly in water:

$$NH_3 + H_2O \longrightarrow NH_4^+ + OH^-.*$$ Since ammonia is a weak base, the ammonium ion produced is a moderately strong acid.

In ammonium chloride solution, there is the base $Cl^-$, as well as the acid $NH_4^+$. But the chloride does not hydrolyze and can be neglected. The ammonium ion hydrolyzes:

$$\underset{\text{Acid}}{NH_4^+} \quad + \underset{\text{Base}}{H_2O} \rightleftharpoons \underset{\text{Base}}{NH_3} \quad + \underset{\text{acid}}{H_3O^+}$$

and the pH of a 0.1M solution of ammonium chloride is a little above 5.

## Hydrolysis: Metal Salts

Salts of most metals hydrolyze fairly strongly to give acidic solutions (the only common exceptions are salts of sodium, potassium, calcium, and magnesium). We'll consider the case of aluminum chloride, $AlCl_3$.

When this salt dissolves in water, it first dissociates to give very weakly basic $Cl^-$ ions, together with highly charged $Al^{+3}$ ions. Each aluminum ion collects a tightly bound cluster of water molecules around it; and it is this aluminum-centered cluster that seems to be the Brønsted acid. See Figure 9-3. An exact formula is not important, but at least three water molecules can take part in a series of reactions, the first of which is:

$$\underset{\text{Acid}}{Al(H_2O)_3^{+3}} \quad + \underset{\text{Base}}{H_2O} \rightleftharpoons \underset{\text{Base}}{Al(H_2O)_2OH^{+2}} \quad + \underset{\text{Acid}}{H_3O^+}.$$

This reaction would be sufficient to account for the acidity of the solution. In dilute solutions of aluminum chloride, two more

---

*The name "ammonium hydroxide" found on bottles in most chemistry laboratories is inaccurate; there is a move on foot to change all these to "ammonia water," but long usage is a stubborn enemy. There are almost certainly no molecules of ammonium hydroxide in ammonia water, and very few ammonium ions and hydroxide ions—about 0.5% in 0.1M solution (i.e., ammonia as a base is almost exactly as strong as acetic acid as acid).

protons are lost to water, creating a fine flocculent precipitate of the base $Al(OH)_3$, still heavily hydrated; i.e., with water molecules still clustered around the aluminum ions of the crystals.

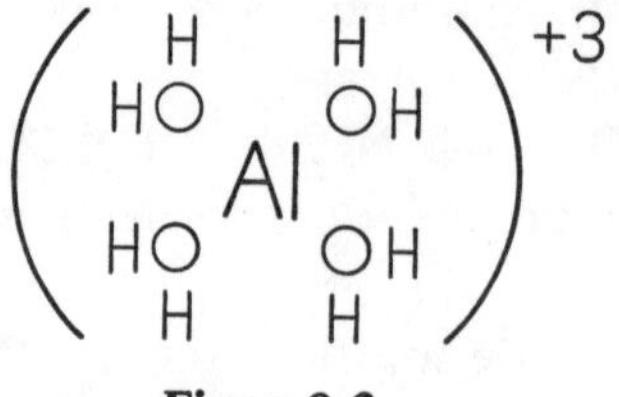

$$\left( \begin{array}{cc} H & H \\ HO & OH \\ & Al \\ HO & OH \\ H & H \end{array} \right)^{+3}$$

**Figure 9-3**

It is interesting that this "base" aluminum hydroxide can *still* act as an acid. It will dissolve in the presence of excess $OH^-$ ions to give "aluminates":

$$Al(OH)_3 \ + OH^- \rightleftharpoons Al(OH)_2O^- \ + \ H_2O.$$

If the source of $OH^-$ ions was sodium hydroxide, the solution would be called sodium aluminate.

*Liquid Ammonia as a Solvent*

One of the great advantages of the Brønsted-Lowry scheme is that it can be extended, unchanged, to solutions other than those in water. Many ionic materials dissolve in liquid ammonia, and acid-base reactions can then occur.

Ammonia as a solvent is surprisingly similar to water. Sketched in Figure 9-4 are models of the two molecules, together with some of the parallels between them when each acts as acid or base.

| Ammonia | Water |
|---|---|
| $NH_3$ + proton→$NH^+$ | $H_2O$ + proton→$H_3O^+$ |
| $NH_3$→$NH_2^-$ + proton | $H_2O$→$OH^-$ + proton |

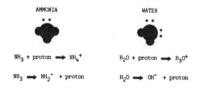

AMMONIA        WATER

$NH_3$ + proton → $NH_4^+$    $H_2O$ + proton → $H_3O^+$

$NH_3$ → $NH_2^-$ + proton    $H_2O$ → $OH^-$ + proton

**Figure 9-4**

Consider now what happens when an acid like HCl is dissolved in ammonia. The reaction is:

$$HCl \quad + \quad NH_3 \longrightarrow CL^- \quad + \quad NH_4^+$$

Acid          Base     Base     Acid.

As with water, the reaction goes far to the right, but it can be reversed by heating (solid ammonium chloride will sublime, and ammonia and hydrogen chloride can be detected in the vapor). The reaction of HCl with water is strongly exothermic, but with liquid ammonia it is almost explosive in its violence. And as the hydronium ion is the strongest acid that can exist in water solution, so the ammonium ion is the strongest acid in liquid ammonia.

In the ammonia system, there is even a parallel to the hydroxide ion; it is the *amide* ion, $NH_2^-$. If sodium amide is dissolved in liquid ammonia, it dissociates, and we have a solution of a strong base. From here on, acid-base reactions will be exactly parallel to those in water. The "neutralization" reaction would be:

$$NH_4^+ \quad + \quad NH_2^- \rightleftharpoons NH_3 \quad\quad + \quad NH_3.$$

Acid(s)     Base(s)     Base(w)          Acid(w)

In the ammonia system an acid like acetic, which is "weak" in water, becomes quite strong. In water it ionizes only slightly, but in ammonia the ionization is nearly complete:

$$HC_2H_3O_2 \quad + \quad NH_3 \rightleftharpoons C_2H_3O_2^- \quad\quad + \quad NH_4^+$$

And the acetate ion that is produced is an exceedingly weak base.

## The Lewis System

The most recent, and the most inclusive, of the acid-base theories is that of G.N. Lewis, who defined an acid as an *electron-accepter* and a base as an *electron-donor*. We cannot go into detail on the Lewis theory here, but it's worthwhile to look at a couple of illustrations.

In the first place Arrhenius' "hydrogen ion" comes back (if only hypothetically), since it would be the all-original electron-accepter; and my example of a lost proton in space finding a water molecule would be an acid-base reaction:

$$H^+ \quad + \quad :\overset{..}{O}:H^- \longrightarrow H:\overset{..}{O}:H.$$

Acid        H          H

          Base

The Lewis system is very effective in explaining *complex* formation, such as the deep blue cuprammonium ion that is produced when ammonia water is added to solutions of copper salts:

$$Cu^{+2} \qquad + 4NH_3 \longrightarrow Cu(NH_3)_4^{+2}.$$

Acid          Base

But it is particularly in organic chemistry that the Lewis system turns out to be a powerful instrument for the prediction of many important reactions.

### "Lime" and the Brønsted-Lowry System

Some of the most ancient knowledge of chemistry surrounds the chemistry of calcium carbonate, "limestone." If limestone is heated in a kiln to very high temperatures, "burnt lime" or "quicklime," calcium oxide, is produced. If the quicklime is "quenched" in water, a tremendous amount of heat is evolved and "slaked lime" results. This, as whitewash or mortar, will slowly combine with carbon dioxide of the air to make limestone again. But limestone can be dissolved by water containing high concentrations of carbon dioxide. This can make limestone caves. And the solution that emerges from this process can again lose $CO_2$ to remake limestone or marble, sometimes in the beautiful formations found in such caverns as the Luray of Virginia.

The chemistry of these processes can be followed in the laboratory, and it is intimately tied up with Brønsted-Lowry acids and bases.

When calcium carbonate is strongly heated, it forms calcium oxide:

$$CaCO_3 \longrightarrow CaO \qquad + \quad CO_2.$$

If the calcium oxide is now added to water, it dissociates, and the oxide ion, a very strong base, hydrolyzes:

$$O^= \qquad + H_2O \longrightarrow OH^- \qquad + OH^-.$$

Base (ss)     Acid(w)   Acid(ww)        Base(s)

Notice that here the hydroxide ion, the strongest base that can exist in water, must be considered an acid! The crystals of calcium hydroxide are called "slaked lime" and are slightly soluble in water, to give a saturated solution that is called "limewater."

It is said that old-time snake-oil peddlers used to have their dupes blow air through limewater, which would presently cloud up. The peddler would then exclaim at all the "disease bugs" in the victim's breath, and sell him some snake-oil to cure him. The limewater reaction consisted of three parts: the formation of a little carbonic acid from $CO_2$ in the victim's breath, then the reaction of this weak acid with the $OH^-$ of the limewater, and finally the precipitation of insoluble calcium carbonate:

$$CO_2 \quad + \quad H_2O \quad \longrightarrow \quad H_2CO_3$$
$$H_2CO_3 + \quad 2OH^- \quad \longrightarrow \quad CO_3^= \quad + \quad 2H_2O$$
$$\text{Acid} \qquad \text{Base} \qquad\qquad \text{Base} \qquad\quad \text{Acid}$$
$$CO_3^= \ + \quad Ca^{++} \quad \longrightarrow \quad CaCO_3 \ (s).$$

You can easily demonstrate this process. But then you can carry it further if you have breath enough (or, easier, use a $CO_2$ generator). If $CO_2$ continues to bubble through the cloudy solution, the cloudiness will clear up completely. The excess $CO_2$ makes enough additional carbonic acid to react with the basic carbonate ions, making bicarbonate ions:

$$CO_3^= \qquad + H_2CO_3 \longrightarrow HCO_3^- \quad + HCO_3^-.$$
$$\text{Base} \qquad\quad \text{Acid} \qquad\quad \text{Acid} \qquad\quad \text{Base}$$

And since calcium bicarbonate is not insoluble, the precipitate dissolves. (Notice again here, that the same ion, $HCO_3^-$, appears twice in the same equation, once as acid and once as base.)

Finally, if the crystal-clear solution is now heated to boiling over a burner flame, a precipitate will reappear, as the last equation is reversed:

$$HCO_3^- \qquad + HCO_3^- \longrightarrow CO_3^= \qquad + H_2CO_3.$$
$$\text{Acid} \qquad\quad \text{Base} \qquad \text{Base} \qquad\qquad \text{Acid}$$

In the hot water, the carbonic acid breaks up to give water and $CO_2$, and the carbonate recombines with the calcium.

It is just these last reactions that occur in limestone caves. Water containing dissolved $CO_2$ under pressure trickles through the calcium carbonate, dissolving it. As this solution reaches regions of lower pressure, the last reaction occurs, and calcium carbonate is slowly redeposited, sometimes in beautiful crystalline forms.

*Experiment*

This experiment takes too long to use as a demonstration, but it is very effective when done by a student.

### Estimation of pH with Indicators

*Introduction*

The pH of a water solution can be measured with high accuracy by electronic instruments, but, for everyday work, colored indicators are used. Each standard indicator has a range of color covering about two pH units. For example bromthymol blue is yellow at pH 6 or below, and changes gradually through green at pH 7, to blue at pH 8 or above. The pH of a solution can easily be estimated to within about half a pH unit by using an appropriate indicator.

In this experiment, we shall first become familiar with the indicator colors and the pH scale by investigating: (a) a series of solutions of a strong acid and (b) a similar series using a strong base. Then in (c) we'll work with solutions of a weak acid, and also of a weak base if desired. Finally, in (d) a series of salt solutions will be tested. The interpretation of the results of this last group presents the greatest difficulty, and the way it can be handled in reporting the experiment will depend on the level of understanding of acid-base theory you have reached.

*Materials*

Solutions of $HCl$, $NaOH$, $HC_2H_3O_2$, and $NH_3$, none of them less than 1M. Also solutions (about 0.1M - 0.2M—the exact concentrations are not very important) of the following salts: $Al_2(SO_4)_3$, $NaCl$, $NaHCO_3$, $NaC_2H_3O_2$, $Na_2CO_3$, $Na_3PO_4$, $NH_4Cl$, $NH_4C_2H_3O_2$—and any others that seem interesting.

Indicators in dropper bottles are provided (with pH range shown) as listed in Table I.

*Procedure*

In general, to test the pH of a solution, put about 4 ml of it in a clean (not necessarily dry) test tube, and add 1-2 drops of the indicator (not more). *Record* the solution, the indicator, the color, and the pH (known or estimated) for that solution. In many cases, two or more indicators may have to be tried before a meaningful result is reached. In each such case, take a fresh sample for each indicator.

*(a) Strong acid.* Start with a solution of HCl of known concentration and prepare a 0.1M solution of it by dilution. For example, dilute 1.0 ml of 3M HCl with water to a final volume of 30 ml. Now use the process of *serial dilution* to make 10 ml each of solutions that are 0.01M, 0.001M, 0.0001M, and 0.00001M. The process: Put 9.0 ml of water in each of a series of test tubes. With a pipette, add 1.0 ml of the 0.1M acid to the first of these

## Table 1†

| Indicator | Approximate pH Range | Color Change |
|-----------|-------------|--------------|
| Thymol blue* | 1.5 - 3 | red - orange - yellow |
| Methyl orange | 3 - 5 | orange - yellow |
| Methyl red | 4 - 6 | red - orange - yellow |
| Bromthymol blue | 6 - 8 | yellow - green - blue |
| Thymol blue* | 8 - 10 | yellow - green - blue |
| Azo violet | 11 - 13 | yellow - brown - purple |

†These indicator solutions can be about 0.1% in water (100 mg in 100 ml). Their preparation requires some care, since most are acids, and not very soluble in water in acid form. Moreover, for careful work, the acidity of the indicator itself can affect results. The procedure I have found useful for preparation of each of these solutions is to weigh 100 mg of the indicator, suspend it in about 50 ml of water, add approximately the amount of 0.10N sodium hydroxide solution shown below, and finally make the volume 100 ml. For high precision, the addition of NaOH should be carried to the point where the indicator is in its "middle" color (e.g., green for bromthymol blue).

| Indicator (100 mg) | Ml of 0.10N NaOH |
|--------------------|------------------|
| Methyl red (hydrochloride) | 4.5 |
| Bromthymol blue | 0.8 |
| Thymol blue (for 8-10 range) | 2.0 |
| Azo violet | 4.0 |
| Methyl orange | 0: (Needs no neutralization) |

Note that the lower range of thymol blue is so acidic that the buffer effect (see Chapter 12) is not significant.

*Thymol blue, shown twice above, has two distinct ranges, both useful. It remains yellow between pH 3 and 8.

tubes, and mix it well. Then take 1.0 ml of *this* solution, and add it to the next tube, etc. You will now have solutions with pH values close to 1, 2, 3, 4, and 5. Use indicators to test samples of each of these (see the general procedure immediately preceding).

Experimental Precaution: The more dilute of these solutions contain so little acid that a tiny amount of impurity in the water or the test tube may invalidate the results, so use distilled (or deionized) water and clean test tubes.

*(b) Strong base.* Do an exactly similar experiment with a series of NaOH solutions, starting with a 0.1M solution and carrying dilutions down to 0.00001M.

A different precaution is in order here. With very dilute solutions of bases, results may be erratic, even though test tubes are clean and the water is "pure." Any water that has stood in contact with air will have absorbed carbon dioxide, giving a very weak solution of carbonic acid (see page 129 of this chapter). The effect of this is ordinarily so small as to be negligible, but when the base being investigated is present in $10^{-5}$ M solution, the carbonic acid can affect the result. For extra-careful work, the water must be boiled to expel carbon dioxide, and then cooled and stored in protected containers.

*(c) Weak acid (and weak base).* Start this time with 1.0M acetic acid solution, $HC_2H_3O_2$, and again make a series of dilutions, at least to 0.0001M (at least five tubes). The pH values of these can *not* be taken directly from the concentration, as with the strong acid. Estimate and record the pH of each solution, within half a pH unit, giving the name and color of the indicator used.

If you wish, do the same thing with a series of tubes with ammonia solution (often labeled $NH_4OH$) over the same concentration range.

We shall outline the calculations needed to determine the pH of these solutions in the chapter on Buffers (Chapter 11).

*(d) Miscellaneous salts.* Now test the various salt solutions previously given in *Materials.* For some of these, you may have no idea in advance where the pH may lie. One method of attack is to test a small sample with bromthymol blue first, to see whether the pH is above or below 7. Then apply new indicators to fresh samples.

*The Report.* If you are to write a full lab report on this experiment, the following notes may be useful. Since parts (a) and (b) are designed primarily for you to learn a technique, the reporting can be very cursory, merely giving the pH, indicator, and color in the form of a table. For part (c) the results will need to be explained, in terms of the material of this chapter. Part (d) will require separate handling for nearly every salt, with an equation for each. There must be an explanation of *why* it is acidic, basic, or neutral. Ionic equations (often called "net" equations) will be needed, and the Brønsted-Lowry treatment should be used.

## 10

## Stoichiometry:

## Basic Chemical Arithmetic

Up to now we have dealt with a good many chemical equations, and with a little chemical arithmetic. But we have been able to avoid putting the two together. A qualitative study of the effect was sufficient where equilibrium shifts occurred in a reversible reaction. We were satisfied with the knowledge that the reaction shifted to the right or to the left.

From here on we'll sometimes have to deal more carefully with quantitative effects, so a short side-trip into chemical arithmetic is needed. The principles involved in calculations of this kind grew out of the concept of the mole, itself a consequence of Avogadro's Hypothesis, discussed in Chapter 1.

For every process in which there is a known chemical change, a chemical equation can be written, and the writing and balancing of equations usually comes early in a chemistry course. The finished equation must represent the chemical *facts*; that is, correct formulas must be shown for the starting materials and for the products, and the numbers of atoms present at the end must exactly equal the numbers present at the start.

Once a balanced equation is in hand, you will find in most textbooks a series of instructions on the handling of "volume-

volume," "weight-weight" problems, etc. Such instructions are sometimes confusing. Yet the problems are easy to deal with in just two categories. Where gases are involved at constant temperature and pressure, Avogadro's Hypothesis in its original form takes care of things. In all other cases, the mole is used as the basis for calculation.

*Gases at Constant Temperature and Pressure*

The first step in this type of calculation (as in every type) is to write a balanced equation. For the sake of adequate exploration, we'll start with a relatively complex reaction—the production of nitric oxide from ammonia, in the Ostwald Process for making nitric acid. The equation is:

$$4NH_3 \quad + 5O_2 \longrightarrow 4NO \quad + 6H_2O.$$

The process is carried out at a temperature and pressure where all the materials are gases. And Avogadro's Hypothesis tells us that if these are all measured at the *same* temperature and pressure, the coefficients in the equation, representing numbers of molecules, are directly proportional to the volumes of the gases. If we were to add together in a reaction vessel 4 liters of ammonia and 5 liters of oxygen, and they reacted completely, we should then find in the vessel 4 liters of nitric oxide and 6 liters of steam.

Let's take a problem based on this reaction. "How many liters of nitric oxide could be obtained from 4.32 liters of ammonia, if all the gases were measured at 600°C, at a pressure of 7.35 atmospheres?" The answer to this would be simply *4.32 liters* of nitric oxide: for every liter of ammonia used, a liter of nitric oxide would be produced. In a way, this is a "trick" question: the *values* given for temperature and pressure are irrelevant, as long as these conditions remain unchanged.

Extend the same problem with the question, "How much oxygen is needed to react with the 4.32 liters of ammonia?" Since 5 moles of oxygen would react with 4 moles of ammonia, then 5/4 (4.32) liters will be needed, or 5.40 liters of oxygen.

This kind of problem can be made more complex. Suppose we ask, "How many liters of nitric oxide could be obtained if 12 liters of ammonia and 10 liters of oxygen react at constant temperature and pressure?" Here the volumes given at the start won't "come out even." One of the gases will be present in excess, and some of it will remain unchanged at the end of the process.

So we must look more closely at the equation. The 12 liters of ammonia that we were given would have to combine with 15 liters of oxygen, for the volume ratio of 5:4 that is required by the reaction. But we are *not given* 15 liters of oxygen. So we must look again. It will be the 10 liters of oxygen that will be used up, and the same 5:4 ratio shows that only 8 liters of ammonia will react with it. In the final mixture, there will remain 4 liters of unused ammonia.

A discussion of this sort is hard to follow. Consequently, for this problem, and for all the others that we'll consider in this and future chapters, the following format for solution will be found useful:

I. Write the balanced equation, spreading it out well on the page.

II. Beneath each formula, write the volume of that material that is given.

III. Follow these, still beneath the appropriate formula, with the amount of that material *used,* and the amount that *remains.* On the right side, show the amount that is *produced.* I usually insist on showing in some very simple way the arithmetical process.

Here is the last problem, presented in this way:

$$4NH_3 \ + \ 5O_2 \longrightarrow \ 4NO \ + \ 6H_2O$$

Given: 12 l            10 l

Used:  $\dfrac{4}{5}$ x 10 = 8 l  10 l

Result: 4 l left        none left        $\dfrac{4}{5}$ x 10 =        $\dfrac{6}{5}$ x 10 =

8 l produced   12 l produced

This scheme presents in orderly form the results of a somewhat complex thought process. Notice that in this case we began with 22 liters of mixed gases, and we finished with 24 liters. These figures would have been incomprehensible to Avogadro's predecessors.

## The General Stoichiometric Problem

The word "stoichiometry" comes from two Greek words that mean element-measurement, and the type of problem just considered is of the simplest type. For all other kinds of problems, the central part of the procedure is to use the *mole* as the basis of calculation.

The mole concept, of course, stems from Avogadro's Hypothesis. Since a *mole of gas* contains a definite number of particles, then when that gas liquefies, or changes to a solid, the number of particles remains unchanged. Therefore, when we speak of a mole, we're talking about a *number of particles,* whether these be atoms, molecules, ions, or whatever. (We'll even use some of the "whatever" when we deal with a colloid; and one of M.J. Sienko's problems asks how many *moles of people* there are on earth.) If molecules react with one another in a fixed ratio, then this same ratio will apply to the number of *moles* that react; and we can use these numbers of moles just as we used gas volumes in the previous class of problem.

The procedure for this type of problem requires four steps.

I. Write the balanced equation, spread out as before.

II. Roughly under each formula, write the quantity given of that material, and immediately under this its *value in moles.* In this conversion, the coefficients of the equation are *not* used: 20 g of CaO is *always* 0.357 moles of CaO, no matter what the CaO is used for.

III. Use the coefficients of the equation to calculate the number of moles (used up or produced) of any other substances in the reaction. The process is just like that used in the all-gas problems above, and requires the simplest of arithmetic—multiplication by an integer or a rational fraction.

IV. Finally, convert moles of product, of excess material, etc., to grams, liters, or whatever is required.

It is wise to keep scratch-work out from underfoot (e.g., calculation of number of grams in a mole of each substance). Multiplications and divisions are conveniently done on a slide-rule, but they should be indicated.

Two examples are shown below, one relatively simple, the second more complicated.*

*Example 1.* How much calcium phosphate can be made from 20

---

*Suggestion for studying. When an *example* is given in this or any other book, first study it until you think you've mastered it. Then shut the book, take a sheet of blank paper—blank except for the statement of the problem—and see if you can do the problem. If not, study it again and repeat. Elementary? Yes, but often neglected.

g of calcium oxide by adding phosphoric acid to it; and how much phosphoric acid will be needed?

I.        $3CaO$    +    $2H_3PO_4$    ——→ $Ca_3(PO_4)_2$ + $3H_2O$

II.    Given: 20 g

$= \dfrac{20}{56}$ = 0.357 m

III.                       Needed: $\dfrac{2}{3}$ (0.357) m  Produced: $\dfrac{1}{3}$ (0.357) m

                                   =0.238  m                    = 0.119  m

IV.                    0.238 x 98 g = 23.3 g    0.119 x 310 g = 36.9 g

                       phosphoric acid           calcium phosphate

*Example 2.* How many grams of iron (III) sulfate will be formed if 7.0 g of iron are added to 15 g of sulfuric acid, and what will be left over? How many grams of it? How many liters of hydrogen gas will be produced (measured at standard temperature and pressure) in the reaction?

I.    $2Fe$    +    $3H_2SO_4$    ——→ $Fe_2(SO_4)_3$ + $3H_2$

II.    $\dfrac{7g}{56}$  $=$  $\dfrac{15 \, g}{98}$  $=$

       0.125 m       0.153 m

Now stop and consider: 0.125 moles of iron would require $\dfrac{3}{2}$(0.125) moles of $H_2SO_4$, or 0.187 moles, for this reaction. But

less $H_2SO_4$ than this is available. Therefore, the $H_2SO_4$ will be used up, and some iron will be left. So we proceed:

III.  Used:  $\dfrac{2}{3}$ (0.153)m      0.153 m

            = 0.102 m

Result:      0.023 m left    0.00 left    $\dfrac{1}{3}$ (0.153) m    0.153 m
                                                                 produced

                                          = 0.051 m
                                          produced

IV.   $0.023 \times 56$                $0.051 \times 400$ g   $0.153 \times 22.4$ l

     = 1.29 g Fe left              = 20.4 g iron    = 3.43 l

                              sulfate             hydrogen gas

## Stoichiometry of Solutions

A good deal of quantitative work in chemistry is done by means of *titration*. A standard solution, containing a precisely known concentration of one reagent, is added from a burette to an unknown solution. The addition continues until the exact stoichiometric amount has been added for the known reaction. This "endpoint" is usually signaled by a color change of an indicator. The volume of the standard solution is read from the burette. And by a simple calculation, the quantity of material in the unknown solution is found. Since the great majority of titrations are of acid against base, I'll deal with this procedure only.

## Normality and Equivalents

A standard solution of an acid is usually labeled in terms of its *normality*. By definition, a 1 N (normal) solution of an acid is one that contains 1 mole of hydronium ion available "on demand" in an acid-base reaction. Thus a 1 N solution of hydrochloric or nitric acid (HCl, $HNO_3$) is simply a 1 M solution, since each of these acids contains 1 mole of protons per mole of acid. But a 1 N solution of sulfuric acid ($H_2SO_4$) is a 0.5 M solution, since each mole of sulfuric acid contains 2 moles of protons.

The words "available on demand" are significant. A solution of acetic acid is only slightly ionized, which means that only a small quantity of hydronium is present in the solution at a given time. But when a solution of acetic acid reacts with a base, the protons are *all* eventually removed from the acid molecules as the neutralization proceeds. So a 1 N solution of acetic acid is also 1 M.

The normality of a basic solution is determined in the same way by the number of available hydroxide ions in the solution. Thus 1 N sodium hydroxide is also 1 M; 1 N Barium hydroxide $(BaOH)_2$) is 0.5M. And 1 N ammonia ($NH_3$) is 1 M, because when a mole of it reacts with acid, 1 mole of protons are taken up.

The term "equivalent" is used for the number of grams of an acid that contain a mole of protons, or the number of grams of base that will furnish a mole of hydroxide ions. So a 1 N solution of acid or base contains 1 equivalent (eq) of available hydronium or hydroxide ions.*

If a specific acid solution is labeled 0.102 N, this means that in a liter of solution, 0.102 moles of hydronium ions are available for neutralization of a base. And a 0.095 N sodium hydroxide solution has 0.095 moles of hydroxide ions in a liter.

*Example:* An unknown basic solution is provided and a drop of indicator is added to it. Hydrochloric acid, 0.102 N, is added from a burette until the indicator just changes color. At this point the burette shows a figure of 24.3 ml. Then the base solution must have contained (0.0243)(0.102) moles of hydroxide ions, or 0.00248 equivalents of base.

In the assaying of ordinary household vinegar, a sample is ordinarily diluted precisely 1:10 by means of standard pipettes and volumetric flasks. Suppose then it is found that 25.0 ml of this diluted acid requires 20.8 ml of 0.0950 N sodium hydroxide solution for neutralization. The calculation here could be done as follows: (0.0208)(0.0950) = 0.00198 eq of acid were present in the 0.0250 l sample. This means 0.0792 eq per liter. The *diluted* vinegar was thus 0.0792 N and the original vinegar 0.792 N. Since vinegar is acetic acid solution, and acetic acid has one acid proton in its molecule, the solution was 0.792 *molar.* A mole of acetic acid weighs 60 g, so the vinegar had (0.792)(60) = 47.5 g of acetic acid per liter. This bottle might have been labeled by the manufacturer as "4.7% acid content."

When the calculation is worked out in the detail I have shown here, it looks very clumsy. In practice it can be much simpler. In the first place, chemists rarely bother to express titration results directly in equivalents, but use the handier unit, *milli*equivalent (meq). Notice that in each of the examples I've used, I multiplied *volume* times *normality* as the first step. If the volume in *milli*liters is used in this process, the product of V x N will come

---

*The same terms, normality and equivalent, are also used in connection with oxidation-reduction reactions, in a sense that is a little more complicated. An equivalent here is the number of grams required to give up, or to receive, 1 *mole of electrons* in a particular reaction. Since reactions of this type can be very complicated, and the same reagent can even have different reactions in different situations, the application of these terms to "redox" reactions is beginning to die out. Such solutions are now often specified in terms of their molarities. As a horrible example, consider iodic acid, $HIO_3$, for which a 1 M solution is 1 N, when used for acid-base work; but this same solution would be 5 N in a reaction where iodine was the final product, and 6 N where iodide ion was produced.

out in milliequivalents. Let's run through the calculations in the two examples again, using this process.

For the first example: 24.3 ml of 0.102 N acid contained $(24.3)(0.102) = 2.48$ meq; therefore, there were 2.48 meq of base in the unknown.

In the vinegar assay, $(20.8)(0.095) = 1.98$ meq of acid were found in the 25 ml of diluted vinegar. This meant 79.2 meq in a liter of the diluted solution, or 792 meq in the *original* vinegar. For the final conversion to grams, we'd use 0.792 moles in the usual way.

For most titrations, indicators are used, as I said, to determine the endpoint: the point where exactly equivalent amounts of acid and base have been added together. Why do we use different indicators for different types of titrations?

The answer is relatively simple, and is directly related to the Brønsted-Lowry acid-base concept. When a strong acid is titrated against a strong base—for example, hydrochloric acid against sodium hydroxide—both reagents are almost completely ionized to begin with, and the only significant reaction is:

$$H_3O^+ + OH^- \longrightarrow 2H_2O.$$ The chloride and sodium ions are merely spectators in the process, and when equimolar amounts of acid and base have been added together, the pH will be exactly 7. Moreover, a tiny amount of excess acid or base will make a very large change in the pH. (Consider that the change from pH 7 to 9 requires the addition of only 0.00001 mole of $OH^-$ ion to a liter of water.) Therefore, any indicator with a color change between 5 and 9 will serve very well to mark the endpoint of the titration.

But when a weak acid, such as acetic, is titrated against sodium hydroxide, the reaction involves acetic acid *molecules:*

$$HC_2H_3O_2 + OH^- \longrightarrow C_2H_3O_2^- + H_2O.$$ The acetate ion produced in this reaction is a moderately strong Brønsted base, and the reaction is therefore reversible, as the acetate ion hydrolyzes (see Chapter 11, page 124). Thus when equimolar amounts of acetic acid and sodium hydroxide have been added together, there is a slight excess of $OH^-$ ion in the solution, with a pH that is between 8 and 9. In this case, an indicator is needed that will change color in this range, rather than at pH 7. The indicator often used is phenolphthalein, which begins to show a pink tinge as the pH reaches about 8.5. I prefer thymol blue (see Chapter 9, page 131) because its color change is more distinct, from yellow to blue, with a clear green at pH 9.

If titrations are done with a weak base and a strong acid, the hydrolysis goes the other way. Ammonia titrations are important when proteins are being measured, and the reaction of ammonia with hydrochloric acid is

$$NH_3 + H_3O^+ \longrightarrow NH_4^+ + H_2O.$$ The ammonium ion, in its turn, is a moderately strong Brønsted acid, leaving some hydronium ion in the solution at the equimolar endpoint. The pH in this case will be in the neighborhood of 5, and methyl red or methyl orange is usually used as an indicator.

I've been throwing pH values around with some abandon in this section, but I've used them merely to show whether a solution is slightly acidic or slightly basic. In the next chapter, we'll have a chance to see how these values are found.

## Problems

A few problems, of increasing difficulty, are given here. They are intended as samples only, to illustrate the material of this chapter. The answers given were worked out on a slide-rule, to about a three-significant-figure accuracy.

1. If 4.8 g of magnesium are burned in air, how many g of product are formed?                                    8.0 g of MgO

2. (Note: The following is a very easy problem. If you find it hard, you're doing it wrong; look at it again and start over.) An excess of sulfuric acid is added to 26 g of zinc, forming hydrogen and zinc sulfate. All the hydrogen so produced is burned in air to form water. Just enough quicklime (CaO) is added to the water to absorb it all, forming solid slaked lime ($Ca(OH)_2$). The latter is dissolved in more water, making limewater, and $CO_2$ is bubbled into it, forming a precipitate of $CaCO_3$ (and of course more water). How many liters of $CO_2$ measured at standard temperature and pressure, will be needed for this last step?      8.96 liters $CO_2$

3. Potassium metal, 3.9 g, are introduced into a flask containing 5.32 g of chlorine gas. After the somewhat spectacular reaction is finished, a white powder, potassium chloride (KCl) remains, together with excess potassium or chlorine. What will be the weight of the latter? And how many *moles* of KCl will be formed?
1.8 g $Cl_2$ ; 0.10 mole KCl

4. 24 g of sulfur and 35 g of iron filings are well mixed together, then heated until they combine (incandescently) forming FeS. The

product in excess is then burned in oxygen (to form either $SO_2$ or $Fe_2O_3$). How many grams of this last oxide are formed?          7.6 g

5. 10.0 g of sodium hydroxide were added to 10 g of phosphoric acid. How much water was produced, if the other product was sodium phosphate?                                                        4.50 g

6. 6.00 g of iron and 6.00 g of sulfuric acid were added together in a flask. Assuming that the hydrogen that was produced escaped, and that iron (II) sulfate was also formed, how much of what materials remained in the flask when the reaction was finished?

2.58g Fe and 9.40 g $FeSO_4$

7. Iron (II) phosphate will dissolve in water that contains sulfuric acid, yielding a solution of iron (II) sulfate and phosphoric acid, but it will not dissolve in non-acidified water. (Note: Water itself does not enter into the calculations.) If 40.0 g of iron (II) phosphate are added to water, will 15.0 ml of pure sulfuric acid (density = 1.80 g/ml) be enough to dissolve it? If not, how many g of the phosphate will remain undissolved? In either event, how many *moles* of iron (II) sulfate will be produced?

No: 7.1 g  undissolved phosphate

0.276 moles iron sulfate

8. Copper metal will dissolve in concentrated sulfuric acid, if it is heated, to yield copper (II) sulfate, sulfur dioxide, and water. How many ml of $SO_2$ will be produced, at 30°C and 740 torr (740 mm Hg), if 5.00 g of copper are added to 15.0 g of sulfuric acid?

1950 ml

9. 25.7 ml of 0.107 N hydrochloric acid neutralized 23.2 ml of a sodium hydroxide solution. What was the normality of the latter?

0.119N

10. A sample of ammonia in water required 36.3 ml of 0.107 N HCl solution to neutralize it to methyl red. How many milligrams of ammonia ($NH_3$) were present in the sample?          66.5 mg

11. 0.206 N sodium hydroxide was used to titrate a sulfuric acid solution of unknown concentration. If a 25.0 ml sample of the acid required 13.6 ml of the hydroxide solution for neutralization, what was the *molarity* of the acid?          0.056 M

12. The labels were lost from three bottles of crystalline materials. It was known, however, that one was salicylic acid (mol. wt. 138), one was benzoic acid (122), and one was potassium hydrogen phthalate (204). Each of these is weakly acidic in water solution, and each has one readily titratable proton. To decide which was

which, 500-mg samples of two of the materials were weighed out, and titrated with 0.100 N sodium hydroxide solution, against thymol blue as indicator. Sample A required 24.5 ml of the base, sample B required 36.2 ml. It was, of course, unnecessary to titrate a sample from bottle C. What labels should be put on the bottles? A is potassium acid phthalate

B is salicylic acid

C is benzoic acid

# 11

## Equilibrium in Buffers:
## How Buffers Work; and How to
## Handle More Advanced Calculations
## in Equilibrium*

Did you ever spill some sulfuric acid on the sleeve of a woolen coat? If so, were you lucky enough to have a box of sodium bicarbonate (baking soda) on hand? If you did, you may have saved the sleeve by rubbing the soda into it. Soda "neutralizes" the acid and you don't have to be careful about how much you use. Any extra soda can either be washed away or allowed to dry and then brushed off.

Of course you could have neutralized the acid with sodium hydroxide solution. Try that on a woolen sleeve some time! No, don't—if you use too little sodium hydroxide, the acid will eat your sleeve; whereas, if you use too much, the alkali will be just as bad. The base will neutralize the acid, but what will neutralize the base?

Let's transfer this neutralization process from your sleeve to a

---

*Portions of the material included in this chapter first appeared in the journal *Chemistry*, and are reprinted here by permission of the American Chemical Society.

beaker and follow it in detail. But first, let's start with a somewhat simpler situation.

Make up a 0.1M acetic acid solution (6 grams per liter or 0.6%), and check its pH using a pH meter, if one is available. If not, you can use an indicator such as thymol blue, which is red in strongly acid solutions, yellow in faintly acid or neutral ones, and blue in basic solutions. The acetic acid solution will give a faintly orange tint to the thymol blue, indicating that its pH is just below 3. This means that the hydronium ion concentration, written $[H_3O^+]$ but often abbreviated $[H^+]$, is a little over $10^{-3}$ M.

If you neutralize the acid by slowly adding sodium hydroxide, the very first drop of base will make the pH rise sharply. The indicator will immediately turn yellow. As you continue adding base, the pH will rise more slowly until the amount you have added is equivalent to the acid, mole for mole. Then the pH will jump sharply, above 9, giving a blue color to the indicator. Just a little more base will move the pH up to 11 or 12, into the region where the solution would soon rot a woolen sleeve (Figure 11-1). This is pretty much what you'd expect when adding a strong base to an acid.

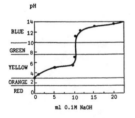 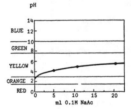

Figure 11-1. Titrating 10 ml 0.1M acetic acid using a
pH meter or thymol blue indicator. Left, with 0.1M
sodium hydroxide; right, 0.1M sodium acetate.

Now start over again with the 0.1M acetic acid solution, but this time add sodium acetate instead of sodium hydroxide. Use the solid salt because you will need a lot of it. Again the pH will rise sharply with the first addition, and the indicator will turn yellow. Obviously, the acid is being neutralized in some way. But you can go on adding sodium acetate until you're tired, without making the indicator turn blue. A pH meter will show that the pH moves up to 4.7 when you've added as many moles of salt as there were originally moles of acid, but then it climbs very slowly (Figure 11-1). Even after you've added 10 moles of the salt for each mole of the acid, the pH will reach only 5.7. It will take a ratio of 100 to 1 to reach pH 6.7, a nearly neutral solution, and you'll have a semi-solid mush long before you reach pH 8.

Why should this be? Obviously, the salt has made the acid much less acidic, but it can't quite make it definitely basic. The clue is in that very first pH reading with acetic acid solution. You prepared a 0.1M solution but the pH reading showed that it was only 0.001M in $H^+$; that is, only about 1% had ionized. Let's see just what that implies.

*Weak Acids*

When you add acetic acid to water, acid and water molecules start bumping into each other. Occasionally they bump in such a way that a proton, $H^+$, is pulled off the acid molecule and attaches itself to the water. Probably, a weak hydrogen bond forms between the two molecules if they collide in the proper position. Then, when they pull apart again in their frantic thermal motion, the proton sometimes sticks with the water instead of with the acetic acid. This is shown here, with the acetate group abbreviated as Ac instead of $C_2H_3O_2$:

$$H\text{--}\underset{\underset{H}{|}}{O} + H\text{--}Ac \longrightarrow H\text{--}\underset{\underset{H}{|}}{O} \ldots H \ldots Ac \longrightarrow$$

$$H\text{--}\underset{\underset{H}{|}}{O}\text{--}H^+ + Ac^-$$

If this proton removal happened very easily, as it does with hydrochloric acid, the two ions would go on their happy way, attracted toward one another, but not re-exchanging the proton. However, in our case, after a few of these effective collisions, a hydronium ion will bump into another (or the same) acetate ion. This collision is pretty sure to work to the disadvantage of the hydronium ion. The hydrogen bond will form again, and the acetate ion will regain its partner (Figure 11-2):

$$H\text{--}\underset{\underset{H}{|}}{O}\text{--}H^+ + Ac^- \longrightarrow H\text{--}\underset{\underset{H}{|}}{O} + H\text{--}Ac.$$

Figure 11-2. $H^+$ and $Ac^-$ form a pair.

The acetate ion is a little like a jealous husband at a dance (Figure 11-3)—sure, his pretty wife can dance with somebody else for a few minutes, but he wants most of the dances for himself (Figure 11-4). (We won't carry this analogy too far, or we might find ourselves in a very promiscuous social system!)

Figure 11-3. Sometimes in aqueous solution, $H_2O$ pairs off with $H^+$.

The effect of this exchange and re-exchange of partners is that the numbers become stable very quickly. At any given time, only a few of the acetic acid molecules have lost their protons. However, the exchange continues constantly—and we have a dynamic equilibrium in which the two reactions are in balance:

$$H_2O + HAc \rightleftharpoons H_3O^+ + Ac^-.$$

Figure 11-4. $Ac^-$ jealously interferes.

This one simple equilibrium will concern us throughout most of the rest of this chapter.

### Acid Plus Salt

Now let's start adding solid sodium acetate to the acetic acid solution. As soon as the crystals hit the water, sodium and acetate ions separate and remain happy as long as there is an oppositely charged ion somewhere close by. From now on we won't worry about the sodium ions—they are merely spectators in the proceedings.

The acetate ions profoundly affect the equilibrium in the acetic acid solution. There are three ways of looking at this: the Le Chatelier approach, the collision approach, and the mathematical approach based on the collision concept.

Le Chatelier's Principle says that if a stress is applied to a system in equilibrium, the equilibrium must change in such a way as to decrease the stress. The partner-swapping of the acetic acid and water has already provided a few acetate ions, and now we stress the system by dumping in still more acetates. The system must cut down their concentration. See Figure 11-5. There is a simple way to do this—by letting them react with hydronium ions and retire to the sidelines (to the left sideline) as part of HAc molecules. Although this does reduce the acetate concentration, as required, it also decreases the hydronium concentration at the same time. Because there weren't many hydroniums there in the first

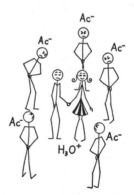

Figure 11-5. When Ac⁻ population is increased, $H^+$ and $H_2O$ still pair off but are separated more rapidly.

place, the effect on them is more dramatic than the effect on the acetates, and the acidity drops sharply; that is, there is a sharp rise in pH.

The situation can also be followed in terms of collisions. Whenever a hydronium ion formed in the original acetic acid solution, it had a relatively slim chance of colliding with an acetate ion, because there was only about 0.001 mole of these in a liter of solution. But then we added acetate ions in the form of the solid sodium acetate and greatly increased the chance that a hydronium ion would find an acetate to bump into. Although this didn't affect the rate of ionization of the acid (which depends only on the number of collisions between acid molecules and water), it did increase the rate of the recombination reaction. Thus, many fewer hydronium ions were present at any given time, and the pH rose.

*Ionization Constant*

Finally, we can consider the "how much" of the process. I don't much care about following someone else's mathematical processes, especially if they're complicated. Assuming you feel the same way, emphasis here will be on simplifying the math. Each of the two reactions we've discussed has a rate which is proportional to the concentrations of the two ions involved (see Chapter 8): $H_2O + HAc \rightarrow$ has a rate which we can write as $K_1 [H_2O] [HAc]$, and $\leftarrow H_3O^+ + Ac^-$ has a rate equal to $K_2 [H_3O^+] [Ac^-]$. Each constant depends on the probability of an effective collision between the molecules (or ions). If the system is in equilibrium, the two rates are obviously equal, or:

$K_2 [H_3O^+] [Ac^-] = K_1 [H_2O] [HAc]$. Easy algebra changes this to:

$$\frac{[H_3O^+] [Ac^-]}{[H_2O] [HAc]} = \frac{K_1}{K_2} .$$

Here, it is a useful convention to put the materials that are on the right-hand side of the equation on the top of the fraction.

Now we start some very practical simplifications. The theories we're using hold with reasonable accuracy only in rather dilute solutions (that's why I chose to start with a 0.1M acid solution). In such solutions, the concentration of water is always about the same, about a kilogram of water (55.5 moles) per liter of the solution—that is, $[H_2O]$ is always about 55.5. Thus, in the equilibrium expression above there are not two constants, but three, which we can roll into one:

$$\frac{[H_3O^+]\,[Ac^-]}{HAc} = \frac{K_1 \times 55.5}{K_2}.$$

(This is called the ionization constant of the acid; for acetic acid, it is about $2 \times 10^{-5}$.)

The size of the ionization constant tells us at a glance a good deal about the strength of the acid. If it is very small, for instance $10^{-10}$, the acid is very weak indeed—boric acid is in this range. Citric acid in lemon juice is about as strong as the acetic acid of vinegar. Phosphoric acid has a constant of about $10^{-2}$ (for the first stage of ionization) and, thus, is very much stronger. The benchtop strong acids, sulfuric, hydrochloric, and nitric, have constants too large to come under this scheme at all.

*Calculation*

We can use the equilibrium expression in a number of ways. First, it can give us a value for the pH of a given solution of a pure acid. Let's run through a calculation for the 0.1M acetic acid we started with, but do it the way a chemist who hates calculations would do it.

Write the chemical equation, and beneath it put two sets of figures. For the first set, use the starting concentrations of all the reacting species and, for the second set, their concentrations after ionization at equilibrium:

$$H_2O + HAc \rightleftharpoons H_3O^+ + Ac^-.$$

Starting
    concentration:    0.1             0        0
At equilibrium:    0.1 - x         x     x

This says that x moles of acetic acid have reacted with water, and, because each acid molecule handed over one proton to a molecule of water, x moles each of hydronium and acetate ions are formed.

Now we can substitute these figures in the equilibrium expression:

$$\frac{x^2}{(0.1 - x)} = 2 \times 10^{-5}$$

and solve for x. But this would be a quadratic equation, and most chemists don't want to solve quadratic equations if they don't have to. So let's cheat. First, a glance at the ionization constant tells us that x will be quite tiny. In this case $(0.1 - x)$ will be very close to 0.1, so we'll call it 0.1, and see what happens. It may not sound fair, but let's try it. The equation to be solved becomes

$$\frac{x^2}{0.1} = 2 \times 10^{-5},$$

and it's not much work to find that x is about $1.4 \times 10^{-3}$ or the pH is about 2.9.

That's fine, but is it right? There's only one thing to do—see if the result checks in the original equation. The check:

$$\frac{(0.0014)^2}{(0.1 - 0.0014)}$$

should equal $2 \times 10^{-5}$. By actual division, we get $2.0 \times 10^{-5}$ to two significant figures. This shows that the cheating was justified, and so we'll do it again, every chance we get. (But never forget that it is cheating. This treatment is enormously useful, but it will not give valid results if the ionization constant is too big, or if high precision is justified in a given case.)

*Acid Plus Salt*

Now let's do a similar calculation for a solution of acetic acid to which we've added some sodium acetate. Strangely enough, by using the same kind of cheating (perhaps a little more outrageous) this calculation is even easier. First we'll add 0.1 mole of sodium acetate salt to a liter of the 0.1M acid solution, and assume that the volume doesn't change. Again, we'll use two sets of figures, one at the starting point and the second at equilibrium.

$$H_2O + HAc \rightleftharpoons H_3O^+ + Ac^-.$$

| Starting point | | | |
|---|---|---|---|
| Original | 0.1 | 0 | 0 |
| Crystals added | 0.1 | 0 | 0.1 |
| Equilibrium | $0.1 - x$ | x | $0.1 + x$ |

Then, we'll put the final concentrations into the equilibrium expression:

$$\frac{(0.1 + x)(x)}{(0.1 - x)} = 2 \times 10^{-5}.$$

And again we'll cheat by dropping x when it is added to or subtracted from 0.1. This time, though, we know that we can't go very far wrong. The equation now becomes

$$\frac{0.1x}{0.1} = 2 \times 10^{-5}$$

which we can solve at a glance—the hydronium concentration is 2 $\times 10^{-5}$ for a pH of about 4.7.

Now let's use the same kind of simple calculation to see why the first addition of crystalline sodium acetate made the pH jump sharply, but later additions had much smaller effects. (See Figure 11-1.) In our equilibrium expression we've been writing

$$\frac{[H_3O^+][Ac^-]}{[HAc]} = 2 \times 10^{-5}$$

and, in the last calculation, the values used for $[Ac^-]$ and for $[HAc]$ were those in the imaginary situation at the starting point; whereas, the $[H_3O^+]$ value was what we had at the end. But if we do this (justified) cheating, the denominator of the fraction will not change throughout the addition of sodium acetate, and we can direct our attention entirely to the numerator.

As soon as we add some acetate to the tiny amount already there, we increase the value for $[Ac^-]$ and, therefore, the $[H_3O^+]$ must fall correspondingly, to keep that constant actually *constant*.

The calculation that we just ran through involved the same amount of salt as of acid mole for mole. But, suppose we add salt until the acetate concentration is ten times as much as the acid. The calculation is:

$$\frac{1.0x}{0.1} = 2 \times 10^{-5},$$

and the hydronium concentration drops only to 2 x $10^{-6}$ for a pH of 5.7. Even if we add 100 times more salt than acid (about 820 grams), the hydronium concentration will move down only to 2 x $10^{-7}$ or a pH of 6.7 which still is very faintly acidic.

## Buffer Effect

So much for calculations needed to get some idea of the size of the effects we're dealing with. Now let's see why the solution with one-to-one acetic acid and sodium acetate is a particularly effective example of what are called buffer solutions. We'll use a pH meter in the mixed solution, or add a few drops of an indicator such as methyl red, which will be orange at pH 4.7.

If we add a few drops of hydrochloric acid to the solution, the pH hardly budges—the indicator scarcely changes color. Yet if the same amount of hydrochloric acid were added to pure water, it would push the pH way down and turn the indicator bright red. Somehow the HAc-Ac⁻ system has acted as a buffer against the strong acid; that is, it has made it act much less acidic.

It isn't hard to calculate the effect, but calculation isn't really needed: Either Le Chatelier's Principle or the collision theory explains it. In the buffer solution, we had a lot of acetate ions which were bumping into hydronium ions and retiring them to the left side of the equation. By adding hydrochloric acid, we added a lot of extra hydronium ions (enough, perhaps to make pure water strongly acidic), but the acetate ions were just waiting to gobble them up. The result was a solution a little poorer in acetate ions, to be sure, but with only a few more hydronium ions than before the hydrochloric acid was added.

The buffer effect works just as well with strong base. If we drop in some hydroxide ions in the form of a little sodium hydroxide, the buffer has enough acetic acid molecules, a Brønsted-Lowry weak acid, to take care of them. The reaction produces more acetate ions, and we're left with a little less acetic acid. But as long as there are any acetic acid molecules left after the reaction with hydroxide, they can still ionize; and, as our calculations have shown, small changes in quantity of acetic acid and acetate ions don't change the pH very much. Even if we add sodium hydroxide until 90% of the original acetic acid is used up, the solution won't become basic.

A buffer solution, then, is one containing relatively high concentrations of a weak base and a weak acid, in the Brønsted-Lowry sense. If a strong acid is added, the weak base tends to neutralize it. If a strong base is added the weak acid takes care of it.

### Biological Buffers

Buffers are of great importance in biological systems, where pH control is often vital. The pH of human blood must remain within a few tenths of 7.4. The buffer systems we've discussed so far have involved a weak acid and its own conjugate base (the easiest type to deal with mathematically); this is by no means the only possible combination. The buffers in blood include amino acids, their ions, proteins, and many other compounds.

Perhaps the most sensitive of the blood's buffers is the one composed of carbon dioxide and bicarbonate ion. The acid is furnished by carbon dioxide plus water; the combination ionizes as if it were carbonic acid, $H_2CO_3$. (A little carbonic acid probably is present.) The base is of course the bicarbonate ion:

$$\left.\begin{array}{c} CO_2 + H_2O \\ \text{or} \\ H_2CO_3 \end{array}\right\} + H_2O \rightleftharpoons H_3O^+ + HCO_3^-.$$

The equilibrium constant for the reaction here (again considering water concentration as constant) is $3 \times 10^{-7}$. By calculating as we did, with the acetic acid-acetate mixture, we find that when equal concentrations of $CO_2$ and $HCO_3^-$ dissolve in water, the hydronium concentration is $3 \times 10^{-7}$, and the pH is 6.5. More $CO_2$ lowers the pH, less raises it.

But the human body has available plenty of $CO_2$. It is produced by the metabolism of every cell and carried by the blood until it is expelled in the lungs. If too many hydroxide ions appear—for instance, from excess basic amino acids after a heavy meal—the carbonic acid of the buffer is there to take care of them. The reaction shifts to the right as more $CO_2$ reacts and bicarbonate ion concentration of the blood rises a little, but the pH remains essentially constant. If there is a threat of too much acid—for example, if you drink a glass of lemonade or eat a grapefruit— hydronium ions are consumed by the bicarbonate ions of the buffer. This pushes the reaction to the left, forming more $CO_2$, which is quickly "blown off" in your breath.

Different biological systems have different pH requirements. Peptic digestion in the stomach seems to work best at about pH 3; hence, the acid or sour taste of regurgitated material. In the lower intestine a somewhat basic pH is needed, and bile furnishes the bases needed to produce and maintain it. Yeast does its best job of alcoholic fermentation in a weakly acid solution. All of these systems, developed over eons of evolution, rely on the same kind of buffer chemistry that we have explored here.

Oh yes, about that sulfuric acid you spilled on your sleeve at the beginning of the chapter. When you rubbed baking soda on it, you set up one of nature's own buffer systems, "blowing off" the acid harmlessly as $CO_2$ and leaving sodium sulfate, along with the excess sodium bicarbonate, to be brushed off.

## Indicators, Hydrolysis, and Related Problems

In earlier chapters, I have talked about hydrolysis reactions and acid-base indicators, each in a qualitative way. For example, hydrolysis makes a solution of sodium phosphate strongly basic; whereas, sodium acetate solution will be weakly basic. And certain specific indicators are useful in detecting these facts. Now we'll look at the numerical aspects. We find that only the vaguest of lines divides these reactions from those of the buffers we have been dealing with, and much of the mathematics is very similar.

Let's begin with the indicators. An indicator is a substance that shows a marked color change over a specific pH range. Each indicator is a complicated organic compound, usually a weak acid. When the acid loses its proton to yield its conjugate base (its negative ion), an internal shift occurs in the molecule, giving the color change. Every two-color indicator that I know about shifts *toward the blue end* of the spectrum as it changes from its acid to its base form; i.e., red to yellow, red to blue, yellow to blue (cf. Table I, Chap 9, page 131). For this discussion, I'll deal with an indicator that has a yellow-blue change and use the abbreviation HIn for the acid molecule.

This indicator will ionize in water, like any other acid:

$$HIn + H_2O \rightleftharpoons In^- + H_3O^+.$$

acid            base

yellow         blue

From the point of view of Le Chatelier's Principle, it is clear that addition of excess hydronium ion to the solution (by adding some hydrochloric acid, for instance) will cause a shift to the left, decreasing the blue color and increasing the yellow. If hydroxide ion is added, it will remove the hydronium and cause a shift to the right, giving a blue solution.

How much will the shift be? At what pH will it occur?

Suppose this indicator acid has an ionization constant of $10^{-5}$. Then its equilibrium expression is:

$$\frac{[H_3O^+][In^-]}{[HIn]} = 10^{-5}.$$

If we had a good deal of the indicator in solution, we could treat it as a *buffer* and use it to control the pH by varying the relative concentrations of the HIn and In⁻. But when we are using it as an acid-base indicator, we add only minute amounts to the solution. And in that case, it will be the pH that controls the relative concentrations of HIn and In⁻.

Do a simple calculation with the expression above. Suppose the pH of a solution is 6; that is, $[H_3O^+] = 10^{-6}$. And add a drop of the indicator. Substitution in the equilibrium expression gives:

$$\frac{(10^{-6})[In^-]}{[HIn]} = 10^{-5}, \quad \text{or:} \quad \frac{[In^-]}{[HIn]} = 10.$$

This means that there is ten times as much blue-colored *ion* as there is yellow-colored *acid*. If the pH of the solution were 4, a similar calculation would give the value of the ratio as 0.1, and the yellow color would be ten times as strong as the blue. At pH 5 the ratio would be just 1, and the two colors would be equally strong, to give a green color to the solution.

For practical purposes, a solution that has ten parts of blue to one of yellow will look blue; and with the opposite ratio it will look yellow. This accounts for the usual assumption that the "range" of an indicator is about two pH units. The midpoint of the indicator—its green color in this case—is seen when the hydronium ion concentration equals the equilibrium constant for the indicator: when $[H_3O^+] = K_{ind}$. By a logical development, the negative logarithms of both of these are used, and we speak of the pK of the indicator: when the pH = pK, the indicator is at its midpoint.

When a solution containing a base and an indicator is titrated with an acid, two processes are actually going on. The added acid is of course neutralizing the base; but it is also neutralizing the indicator, changing it from its ionic (basic) form to its acidic form. In this process some of the protons of the acid are taken up by the indicator, so more acid is used than if no indicator had been there. This effect is usually negligibly small: a couple of drops of indicator should make a difference of only thousandths of a ml of 0.1N acid. But when very dilute solutions are being titrated (or if very large amounts of indicator were to be used), an "indicator correction" must be made.

Summing up, an indicator is a colored acid whose conjugate base has a different color. The pH value at the middle of the color change is the pK value for the indicator. And the useful pH range is from about one unit below to one unit above the pK value.

*Hydrolysis*

When a negative ion hydrolyzes, the reaction is:

$X^- + H_2O \rightleftharpoons HX + OH^-$. This reaction has an equilibrium constant, as has any other equilibrium reaction, but you won't find the equilibrium constant listed in handbooks. The reason for the omission is that hydrolysis constants are easily derived from the ionization constants of the acids, after some ingenious mathematics.

The equilibrium expression for this hydrolysis reaction is:

$$\frac{[OH^-][HX]}{[X^-][H_2O]} = K$$

or, with the usual assumption of constant water concentration:

$$\frac{[OH^-][HX]}{[X^-]} = K_{hydr.}$$

Since we don't know $K_{hydr}$, we must manipulate the expression to find it. Note first that we *do* know the value of $[H_3O^+][OH^-]$. So we'll arbitrarily introduce the value of $[H_3O^+]$ into the expression, multiplying both numerator and denominator by this value (even though we don't know what it is, numerically!). The process is legitimate in two ways: (a) there *is* hydronium ion in every water solution, and (b) the value of any expression remains unchanged if it is multiplied by unity; in this case, by $\frac{[H_3O^+]}{[H_3O^+]}$.

The result of this mathematical trickery is:

$$K_{hydr} = \frac{[H_3O^+][OH^-][HX]}{[H_3O^+][X^-]}.$$

The expression can now be separated into two manageable factors, both of whose values we know:

$$K_{hydr} = ([H_3O^+][OH^-])(\frac{[HX]}{[H_3O^+][X^-]}).$$

The first part of this is $K_w$ for water, equal to $1.0 \times 10^{-14}$ (see Chapter 9); the second part is the reciprocal of the ionization constant, $K_{diss}$, of the acid HX. This, you remember, was $\dfrac{[H_3O^+][X^-]}{[HX]}$.

Thus we can find the value of $K_{hydr}$ from readily available information:

$$K_{hydr} = \frac{K_w}{K_{diss}} \,.$$

We'll illustrate this by a simple example: Find the pH of 0.1M sodium acetate solution.

When sodium acetate dissolves in water, it dissociates completely to give $Na^+$ ions and $Ac^-$ ions. The sodium ions will not further concern us—they remain spectators. The acetate ions hydrolyze:

$$Ac^- + H_2O \rightleftharpoons HAc + OH^-.$$

The equilibrium expression is:

$$\frac{[HAc][H_3O^+]}{[Ac^-]} = K_{hydr}, \text{ and this, as we saw, is } \frac{K_w}{K_{diss}}.$$

The values of these constants are $1 \times 10^{-14}$ and (for acetic acid) $2 \times 10^{-5}$. Thus $K_{hydr}$ is $5 \times 10^{-10}$.

Now we're ready to proceed in the usual way:

$$Ac^- + H_2O \rightleftharpoons HAc + OH^-$$

Initial: 0.1                 0        0

Final:  0.1 − x              x        x

Substituting: $\dfrac{x^2}{(0.1 - x)} = 2 \times 10^{-10}.$

To solve this we throw out the tiny value of x where it is added to 0.1 (since doing this will not measurably change the value of the denominator) and get the very simple expression, $x^2 = 2 \times 10^{-11}$. Solving this, x turns out to be $4.5 \times 10^{-6}$.

This is the *hydroxide* ion concentration in the final solution. We wanted to find the pH, and we know that

$[H_3O^+][OH^-] = 10^{-14}$. So $[H_3O^+]$ here is $\dfrac{10^{-14}}{4.5 \times 10^{-6}}$, or $2.2 \times 10^{-9}$,

for a pH of 8.7. This is the "slightly basic solution" that was mentioned earlier.

When hydrolysis reactions yield *acidic* solutions, a process is used that is similar to this in every way. Ammonium salts are the easiest to deal with, as well as the most important. We'll use ammonium chloride. In this case the chloride ion is a spectator (since it undergoes negligible hydrolysis itself), and the reaction is:

$$NH_4^+ + H_2O \rightleftharpoons NH_3 + H_3O^+.$$

This time we multiply both numerator and denominator of the equilibrium expression by the *hydroxide* ion concentration:

$$K_{hydr} = \frac{[NH_3][H_3O^+]}{[NH_4^+]} = \frac{[NH_3][H_3O^+][OH^-]}{[NH_4^+][OH^-]} = \left(\frac{[NH_3][H_3O^+]}{[NH_4^+]}\right)[H_3O^+][OH^-] = \frac{K_w}{K_{NH_3}}.$$

If $K_{NH_3}$ is $2 \times 10^{-5}$ (it happens to have the same value as the ionization constant for acetic acid), then the hydrolysis constant is $5 \times 10^{-10}$. And a calculation like the previous one shows that the $[H_3O^+]$ of 0.1M ammonium chloride solution is $4.5 \times 10^{-6}$, giving a pH of about 5.3, a slightly acidic solution.

I mentioned all these reactions, in a qualitative way, in Chapter 9, and there I said that solutions of sodium phosphate and sodium sulfide are strongly hydrolyzed. It is interesting to work out the pH of these solutions at 0.1M concentration—a job I'll leave to you. The hydrolysis reactions are:

$$PO_4^\equiv + H_2O \rightleftharpoons HPO_4^= + OH^- \quad K_{HPO_4^=} = 10^{-12}$$
$$CO_3^= + H_2O \rightleftharpoons HCO_3^- + OH^- \quad K_{HCO_3^-} = 5 \times 10^{-11}.$$

In both of these cases the mathematical "cheating" that I mentioned earlier in this chapter *cannot* be used, since the value of the final hydroxide concentration is far from tiny. In such a case, the quadratic equation must be solved in the normal algebraic way.

## What Indicator Do I Use?

When acetic acid is titrated with sodium hydroxide, we must use an indicator that changes color when the number of moles of added base exactly equals the number of moles of acid originally present. This seems obvious. But notice that at this point the solution is *not* exactly neutral. We have just calculated that a 0.1M solution of "sodium acetate" actually contains enough hydroxide ions to make the pH about 8.7, along with some non-ionized acetic acid. Thus we want an indicator that will show a clear color change in this region, not at "true neutrality." This is the reason for using phenolphthalein or thymol blue. The pK for each of these is close to 8, and each will show its full basic color at pH 9, where all the acetic acid has been titrated.

On the other hand, when ammonia is titrated with hydrochloric acid (as is done in various kinds of nitrogen analysis), the finished solution of ammonium chloride will have a pH of about 5.3, as we just saw. Consequently, methyl red, or any other indicator with a pK of about 5, would be ideal. Methyl orange, with a pK of about 4, is often used, with insignificant error.

## Titration with a pH Meter

Electronic pH meters are now relatively inexpensive, and they are sometimes used in practical work for titration. In general, the use of indicators is still much simpler. But when colored or cloudy solutions are being handled, indicators may be worthless. The following experiment covers the important aspects of titration with a pH meter. I give here all the instructions (including the introduction) for the experiment as it is used in my classes.

## Introduction

During an acid-base titration, there is a change of hydronium ion concentration. In practical work, we are concerned primarily with the point in the titration where the molar concentration of acid and base are equal—the neutralization point or "endpoint." However, except in the simplest cases (strong acid vs strong base), this point is rarely at exact neutrality in the proper sense of the term. True neutrality is at pH 7.0. At the endpoint of a titration, there is always a rapid change of pH.

When a *strong* acid is titrated against a *weak* base, or vice versa, the salt formed on completion of the reaction will hydrolyze, giving an acidic or basic solution. There will still be a rapid change of pH at the endpoint, but it will occur at some point other than pH 7.

In this experiment, we shall follow, and graph, the changing pH when (a) a strong acid is titrated against a strong base; (b) when either a weak acid or a weak base is titrated against a strong base or strong acid, respectively. Then we shall see what happens when a polyprotic acid is titrated with a strong base. The experiment may be extended at will, to cover highly colored solutions, foods, etc.

## Equipment

A 50-ml burette. A pH meter. Common lab glassware. Standardized solutions of NaOH (N/10) and of HCl (N/10) (the exact concentration of each of these should be known, but they need not be of precisely the same concentration). N/10 solutions (approximately) of ammonia (ammonium hydroxide), and of acetic acid. M/10 (not N/10) solution of phosphoric acid.

## Procedure

*I. Strong Acid vs Strong Base.* Measure 10.00 ml of the standard acid into a beaker of about 150-ml capacity, and add about 80 ml of water. (The purpose of this added water is to make the volume about 100 ml ±10%, throughout the experiment.) Lower the electrode(s) of the pH meter into the solution. If the meter has been properly standardized, it should read about pH 2, since the solution is approximately 0.01N acid. If the reading differs from 2 by less than 0.3 pH units, the titration may proceed. Otherwise, restandardize the meter.

Now add NaOH solution from a burette, 1 ml at a time, for the first 9 ml, followed by additions of 0.2 to 0.3 ml at a time until the sudden pH change. After each addition, swirl or stir the solution (caution: the probe of the pH meter is delicate) and read the pH meter and the burette, recording all readings. The important items are the burette reading and the pH; it doesn't matter if more, or less, than 1 ml of base is added for a given reading. After the sudden pH change, additions of 1 ml at a time (or even more) can be made. Continue until 20 ml of NaOH solution have been added. The data will give a characteristic curve when pH is plotted (y-axis) against ml of NaOH (x-axis).

*II. Strong Base Against Weak Acid.* Measure 10 ml of N/10 acetic acid into a beaker, and add 80 ml of water, as before. Titrate in exactly the same way, until 20 ml of base have been added. Again a characteristic curve results, which is distinctly different in some parts from the strong-acid, strong-base curve (but identical in others). If desired, this titration curve may be superimposed on the one from Procedure I, to bring out similarities and differences. When this is done, it is a good idea to use different colors for the two curves.

*III. Weak Base Against Strong Acid.* Use 10 ml of N/10 ammonia solution instead of acetic acid and add water as before; this time use standard HCl in the burette.

*IV. Polyprotic Acid Against Strong Base.* Start this time with 10 ml of about 1/10 *molar* phosphoric acid solution. Make the small additions of NaOH solution between 9 and 11 ml, and between 19 and 21 ml. And carry the NaOH addition on to at least 31 ml.

### The Report

Describe briefly all parts of the experiment that you have done. Include graphs of the curves; explain briefly why they have the shapes they have, and why they differ.

### Optional

*Quantitative Titration of a Colored Solution.* You will be given a solution of unknown concentration (except that the concentration will be somewhere near N/10), which may be a weak or strong acid or base or a polyprotic acid. You will try to determine which of these it is, and what its exact concentration is.

First determine whether the unknown solution is acidic or basic, then proceed as you did in the main part of the experiment. The principal difference here is that you will have no idea where the pH "jump" will come. Therefore, it is worthwhile to do a first trial without great care, to see at *about* what volume this jump occurs. Now take a sample of the same size, precisely measured, and add standard solution until you are within 1 ml or so of the probable endpoint. Then titrate much more slowly (perhaps a drop at a time at the crucial point) to find the point of maximum change of pH. Carry on to at least 1 ml beyond this point, to establish the shape of the curve. This procedure should be repeated a second time, with a new sample of the unknown. Use the data as described below.

Try a sample of orange juice, diluted appropriately (you may have to experiment a little to find the appropriate dilution). The principle acid here is citric acid, which is polyprotic, but with a first dissociation constant of $8 \times 10^{-4}$.

For the *report* on these titrations, give the data table for the precise titration in each case, and in making the graph, use an expanded scale in the neighborhood of the endpoint (let 1 ml be at least ten visible units on the x-axis). From this graph, read off, to 0.01 ml, the endpoint, and from this calculate the concentration of the unknown solution. From the shape of the curve, and the pH of the endpoint, determine whether the acid or base was weak or strong, whether polyprotic, etc.

## Problems

Here again are a few problems that illustrate some of the points brought up in this chapter.

1. Find the hydroxide ion concentration, and the pH, of a 0.01 M solution of ammonia ($K = 2 \times 10^{-5}$). $[OH^-] = 4.5 \times 10^{-4}$; pH = 10.7
2. Find the pH of a buffer solution made by adding 0.2 moles of acetic acid to a liter of 0.1 M sodium acetate solution. ($K_{diss}$ for acetic acid = $2 \times 10^{-5}$).                                      pH = 4.4
3. Find the hydronium ion concentration, and the pH, of a 0.1 M solution of carbonic acid. Note that although carbonic acid is dibasic, the second ionization can be neglected, since the hydronium ion produced in the *first* ionization will practically preclude further ionization. $K_1 = 3 \times 10^{-7}$.        $[H_3O^+] = 1.7 \times 10^{-4}$; pH = 3.8
4. What is the pH of a solution made by adding 250 ml of 0.1M sodium hydroxide to 750 ml of 0.1 M acetic acid? (Note: This problem can be done by either a very complicated or a very simple method. The simple method is to assume that as a first step *all* the hydroxide ion reacts with acetic acid to produce acetate ion. Then use the resulting concentrations of acid and ion to calculate the final pH.)                           $[H_3O^+] = 4 \times 10^{-5}$; pH = 4.4
5. What is the pH of a solution made by adding 250 ml of 0.2 M hydrochloric acid to 750 ml of 0.1 M sodium acetate solution? The procedure here is exactly similar to that in Problem 4. And the answer is the same, because the final solutions are the same! You may notice a strange similarity to the answer of Problem 2.
6. Find the pH of a 0.2 M solution of $KH_2PO_4$, assuming that the $H_2PO_4^-$ ion is an acid with a dissociation constant of $6 \times 10^{-8}$
$$pH = 8.0$$
7. Find the pH of a 0.1M solution of $Na_2HPO_4$. In this case the negative ion is a strong enough *base* (even though it contains a proton), so the salt hydrolyzes, to yield $OH^-$ ions, and the weak acid $H_2PO_4^-$, with the same constant as in Problem 6.    pH = 10.1

## 12

## The Solvay Process Teaches

## Uses of Dynamic Equilibrium in Industry*

Most students in an elementary chemistry course would be able to complete the equation, $Na_2CO_3(aq) + CaCl_2(aq) \longrightarrow$ by supplying the products, $CaCO_3(s) + 2NaCl(aq)$. This reaction goes far to the right because calcium carbonate is insoluble in water. The constant for the reaction would be about $10^8$.

However, this reaction is of little commercial interest. The sodium carbonate used at the start is a valuable starting material for all kinds of large-scale processes—the soap and glass industries alone consume millions of tons. On the other hand, the calcium carbonate and sodium chloride products are almost dirt cheap. Calcium carbonate can be quarried in great chunks from any limestone deposit, and you can get sodium chloride from salt mines or by evaporating seawater.

### The "Impossible" Process

But reversing this equation, taking two of the cheapest available materials and producing valuable sodium carbonate, is something

*Portions of the material included in this chapter first appeared in the journal *Chemistry*, and are reprinted here by permission of the American Chemical Society.

else again. It looks impossible, but it is just what the Solvay Process does. We shall examine the process in detail, not because it is commercially important (although it certainly is), but because it involves ingenious use of equilibrium shifts, including the common ion effect, careful consideration of equilibrium constants, and the Brønsted-Lowry acid-base system. (See Figure 12-1).

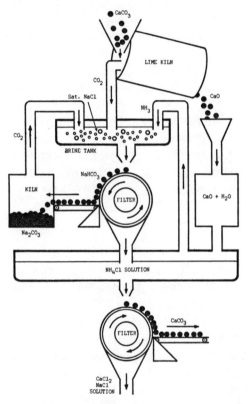

Figure 12-1. The Solvay Process. Calcium carbonate and sodium chloride are converted into sodium carbonate. Ammonia used in the process is recovered and used over.

The process is begun with one big energy input—limestone is "burned" (the old term for it) forming calcium oxide and carbon dioxide:

$$CaCO_3(s) \longrightarrow CaO(s) + CO_2(g) \tag{1}$$

This reaction goes to the left at low temperatures; but, if enough heat is added, it goes to the right. It is one of the oldest known chemical reactions; in fact, calcium oxide, or quicklime, was used to make mortar for the Coliseum in ancient Rome.

The next step is deceptively simple: Ammonia and carbon dioxide from Equation 1 are bubbled into a cold saturated solution of sodium chloride. Considerable heat is evolved in this reaction (or series of reactions) so it must be well cooled. The two gases combine with water, with one another, and with sodium ions of the solution, and sodium bicarbonate (baking soda) precipitates. Then, the sodium bicarbonate is filtered and heated to give sodium carbonate. And there we are! It sounds (and is) so easy—carbon dioxide plus ammonia plus salt and presto—sodium bicarbonate! True, there is one little practical consideration. Where does the ammonia come from? Ammonia is comparatively expensive, and, if it were used up in the process, we'd be out of luck, because the ammonia we put in would cost more than the value of the sodium carbonate we got out. However the quicklime from Equation 1 handles this, and we'll see how later.

### Brønsted-Lowry Reactions

Now we come to the point of this chapter—the reactions that occur in the cold salt water. First, ammonia gas dissolves very readily in water, forming what is labeled ammonium hydroxide on most lab benches. The label is really inaccurate, because most of the ammonia is in solution as ammonia molecules, $NH_3$. The solubility is high because both ammonia and water are highly polar, and, as a result, hydrogen bonds are established between them. Thus, we can get a lot of ammonia into solution to take care of the next step.

Carbon dioxide, the other gas that we're piping in, is not particularly soluble, except at high pressures like those used to make soda pop, beer, or champagne; however, the low temperature of the salt solution increases solubility a little. The small amount that does dissolve probably reacts to a small extent with water to form carbonic acid (chemists tend to hedge on whether this acid exists, but they treat it as though it does). As an acid, it would ionize first to give bicarbonate ions:

$$H_2CO_3 + H_2O \rightleftharpoons HCO_3^- + H_3O^+ \qquad (2)$$

with an equilibrium constant, $K = 3 \times 10^{-7}$. Then it could ionize again, but to a lesser extent to give:

$$HCO_3^- + H_2O \rightleftharpoons CO_3^= + H_3O^+ \tag{3}$$

with a much smaller constant, $K = 6 \times 10^{-11}$. The size of these equilibrium constants is important as we shall see.

As ammonia and carbon dioxide gases are piped into the cold salt solution, the first major reaction is exchange of a proton to make high concentrations of bicarbonate and ammonium ions:

$$H_2CO_3 + NH_3 \rightleftharpoons HCO_3^- + NH_4^+ \tag{4}$$

This is an acid-base reaction, and, according to the Brønsted-Lowry system, the products must also be an acid and a base. In this case bicarbonate ion is a base, as it was in Equation 2 (in Equation 3 it was an acid), and ammonium ion is an acid. The bicarbonate ion is a much weaker base than ammonia gas—their ionization constants in the reaction with water to produce hydroxide ion differ by a factor of about 1000. Also, the ammonium ion is a weaker acid than carbonic acid by about the same factor. Thus, the reaction proceeds quite far toward the right, with a constant on the order of 500.

The reaction immediately uses up the small amount of $H_2CO_3$, and, by a cascading of Le Chatelier effects, allows a very large amount of $CO_2$ to dissolve. Disappearance of $H_2CO_3$ shifts the $CO_2(aq) \rightleftharpoons H_2CO_3$ to the right; and the consequent disappearance of $CO_2$ in solution shifts the $CO_2(g) \rightleftharpoons CO_2(aq)$ to the right.

While we're dealing with this part of the process, we must consider another question. Why doesn't the ammonia push the reaction one stage further, and produce carbonate ions according to the equation:

$$NH_3 + HCO_3^- \rightleftharpoons NH_4^+ + CO_3^= \tag{5}$$

Again the Brønsted-Lowry system helps answer the question. As before, ammonia would act as a base, but this time bicarbonate ion would be an acid. The acidic reaction product would be ammonium ion as before, and the basic product would be carbonate ion.

However, the carbonate ion on the right is a little stronger base than the ammonia on the left—their ionization constants differ by a factor of about 10. Also, the ammonium ion is a little stronger acid than the bicarbonate ion by about the same factor. The equilibrium constant for the whole reaction is about 0.1, and the reaction has practically no tendency to compete with Reaction 4.

It turns out that ammonia is the ideal base for converting carbon dioxide to bicarbonate ions.

## Precipitation

So far I've spoken as if the reactions between these gases were taking place in pure water. But remember, the solution they were bubbled into was saturated with sodium chloride, and was very cold. Salt is nearly as soluble in cold water as in hot—about 35 grams of salt will dissolve in 100 grams of water.

Thus, we have achieved a solution with rather high concentrations of sodium and bicarbonate ions which finally becomes supersaturated with respect to sodium bicarbonate. Consequently, the bicarbonate precipitates:

$$Na^+(aq) + HCO_3^-(aq) \rightleftharpoons NaHCO_3(s) \qquad (6)$$

Sodium bicarbonate is relatively soluble in cold water (about 7 grams per 100 grams of water), but precipitation is enhanced by the common ion effect, because there is excess sodium ion from the salt (see Chapter 8). Finally, about two-thirds of the sodium ions in the original salt appear in the sodium bicarbonate.

To make the sodium carbonate product, the bicarbonate is filtered off and heated:

$$2NaHCO_3(s) \rightleftharpoons Na_2CO_3(s) +$$
$$CO_2(g) + H_2O(g) \qquad (7)$$

Carbon dioxide is returned to the process by piping it back to be combined with the gas from the limestone in Equation 4, and we have achieved our objective, sodium carbonate.

It would be nice, of course, if we could get sodium carbonate directly by adding the two gases to the salt water, but there are two excellent reasons why this won't work. First, sodium carbonate is quite water soluble, and wouldn't precipitate even if it did form, and second, no appreciable amount of carbonate ion is present (as mentioned in discussing the hypothetical Equation 5).

## Ammonia Recovery

But wait a minute! What about the ammonia? As I said earlier, this is relatively expensive and the process would be almost worthless if it stopped here, because the ammonia would cost more than the "soda ash" that was produced.

Here is where the cycle is finally completed. Remember the quicklime, CaO, formed in Equation 1? The oxide ion is a powerful Brønsted base. And in Equation 4, we made a lot of ammonium ion that seemed to be ignored in discussing the bicarbonate ion. To make the situation a little clearer, let's rewrite Equation 6, this time including the "spectator" ($NH_4^+$ and $Cl^-$) ions:

$$Na^+ + Cl^- + NH_4^+ + HCO_3^- \rightleftharpoons$$
$$NaHCO_3(s) + NH_4^+ + Cl^- \qquad (6a)$$

After sodium bicarbonate was filtered off, these spectators remained in the filtrate along with unused salt.

Now if we dump quicklime into the solution, first mixing it with water if desired, and then warm it a bit if necessary, out comes the ammonia. Let's handle this in two stages, the first being the slaking of quicklime with water:

$$CaO(s) + H_2O \rightleftharpoons Ca(OH)_2(s) \qquad (8)$$

This is one of the few almost irreversible reactions in this whole process: the oxide ion is a very strong Brønsted base.

Slaked lime is only very slightly soluble in water:

$$Ca(OH)_2(s) \rightleftharpoons Ca^{++}(aq) + 2OH^-(aq) \qquad (9)$$

Then, the hydroxide ion (strong base) reacts with the acidic ammonium ion:

$$OH^- + NH_4^+ \rightleftharpoons NH_3 + H_2O \qquad (10)$$

This removes the hydroxide ion as fast as it can dissolve, until all the slaked lime goes into solution and reacts.

Ammonia gas is driven out of the solution by heat, some of which is provided by the slaking of the lime. Thus, valuable ammonia is returned to the process at Step 4, and we are left with a solution containing primarily the spectator ions, $Ca^{++}$ and $Cl^-$, and excess sodium chloride left behind in Equation 6.

To tie up one loose end, the bicarbonate ion that did not precipitate in Equation 6 reacted with the hydroxide ion of the lime to make water and carbonate ions which precipitate as calcium carbonate:

$$HCO_3^- + OH^- \longrightarrow CO_3^= + H_2O \qquad (11)$$
$$CO_3^=(aq) + Ca^{++}(aq) \longrightarrow CaCO_3(s) \qquad (12)$$

The precipitate is a very fine powder, which is recovered, and is of commercial value. It is known as precipitated chalk and is used for making such products as blackboard chalk, whiting, and some silver polishes.

We have now succeeded in reversing the equation put to our beginning chemistry student. We consumed only limestone and salt, and produced sodium carbonate and calcium chloride. Most of the energy was added in the very first step, the burning of the limestone. From there on, the energy ran mostly downward, with some heat being evolved in Equations 4 and 8, and a little more being added at Equations 7 and 10. The rest of the reactions occurred because of a cascading of Le Chatelier effects.

If you know of any use for millions of tons of calcium chloride containing some salt, the nation's soda manufacturers would be very grateful to you. They can get rid of trifling amounts, such as a few million tons here and there for making fast-setting cement or for keeping down dust on dirt roads, but they literally don't know what to do with the bulk of the stuff.

# 13

## Analytical Processes: Equilibrium
## Shows Why They Are Effective

One of the main jobs of the chemist, ever since there have been chemists, has been analysis: finding out what was in a material (qualitative analysis) and how much was there (quantitative). Analytical methods depend on all sorts of tricks—ingenious applications of theory. One standard procedure is to form a precipitate, which can be filtered out and identified (qualitative) or measured (quantitative). When a precipitate is ionic, the principles involved in its formation are ones that we have already considered in earlier chapters here.

### The Chloride Test

At some point in every first-year chemistry course, the "chloride test" is introduced. It normally rates no more than a short paragraph in the textbook, and its delightful intricacies are rarely given their due.

The test itself is simple: *Step I:* Add some silver nitrate to the solution being tested; if a white precipitate forms, you may *suspect* that a chloride is present. *Step II:* Now add to the mixture ammonia water in excess, and stir it ("ammonium hydroxide"; see Chapter 11). If the precipitate dissolves, it *may* have been silver chloride. *Step III:* To the clear solution add enough nitric acid to neutralize the ammonia, and then a little more. If a precipitate reappears, and remains, it *must* be silver chloride.

There are many negative ions that will give a precipitate with silver in Step I. A few of these precipitates will not dissolve in ammonia in Step II: the iodide and bromide are two of these, and are eliminated at this stage. For the rest, when the ammonia is neutralized with nitric acid in Step III, some of the original precipitates will reappear fleetingly, but all of these except the chloride will redissolve in excess nitric acid.

For a demonstration of this test, line up a series of test tubes, each one containing a dilute solution of the sodium or potassium salt of interesting negative ions. A good selection would include iodide, bromide, chloride, carbonate, phosphate, and chromate. See Figure 13-1.

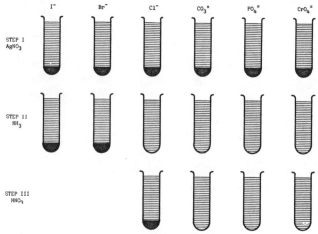

Figure 13-1. Observed changes in the three steps of the "chloride test."

Add a drop or two of silver nitrate solution to each of the test tubes, and shake them. A precipitate will appear in each. Some of these precipitates will be colored; the chromate, for instance, is a bright maroon. Now add a good portion of ammonia to each tube, and shake or swirl it. All the precipitates will dissolve except the bromide and the iodide (with enough ammonia the bromide *may* dissolve, but a large overdose of concentrated ammonia is needed). These two tubes, that still have precipitates, are set aside as non-chlorides.

Nitric acid is now added, drop by drop, to each of the other tubes in turn, with mixing. In the tube with the carbonate, nothing is seen except a little fizzing. With the phosphate and the chromate, the original precipitate will show up again when the ammonia has been just neutralized; then as more acid is added, the solutions will clear completely. But in the chloride tube, the precipitate will reappear and remain when excess acid is added.

### What Happens with Chloride Ions?

This whole demonstration involves four distinct kinds of equilibrium. We'll start with those that occur in the chloride tube only. Throughout this discussion we'll ignore the sodium, potassium, and nitrate ions, which take no part in any of the reactions—they are merely "spectators."

On adding the silver nitrate, in Step I, the "insoluble" salt, silver chloride, precipitates (Figure 13-2):

$$(1) \quad Ag^+ + Cl^- \longrightarrow AgCl(s).$$

But this reaction, like every chemical reaction, is reversible. The precipitate is in equilibrium with tiny amounts of ions in solution. These amounts are *very* tiny, since the solubility product of silver chloride is only about $2 \times 10^{-10}$. Still, at every instant, some of the solid is dissolving, at the same time as ions from the solution are attaching themselves to the crystals of precipitate (cf. Chapter 8).

**Figure 13-2**

The silver ion, like the ions of many other transition elements, can form "complexes" with ammonia. The metal ion is joined to ammonia molecules by weak covalent bonds (Figure 13-3):

$$(2) \ Ag^+ + 2NH_3 \rightleftharpoons Ag(NH_3)_2{}^+.$$

This silver complex is soluble in water, but the normal properties of the silver ion are "blanketed" by the ammonia molecules. The equilibrium constant for this reaction is about $10^{-7}$, which means that it goes far to the right.

When in Step II ammonia is added to the mixture containing the precipitated silver chloride, it immediately combines with the few silver ions present in the solution. These are therefore no longer available to combine with the chloride ions, and the precipitation of reaction (1) can no longer happen. But that precipitate is still dissolving, as it has been doing all along; and as each ion-pair of silver and chloride goes into solution, the ammonia gobbles up the silver ion (or, less drastically in my sketch (Figure 13-3), grabs it). Soon, if enough ammonia is present, the precipitate has all dissolved. There are two *competing* equilibria. The silver may combine either with chloride or with ammonia, and the ammonia combination (which is soluble) wins the competition.

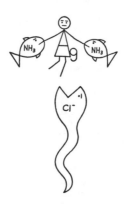

Figure 13-3

But now, in Step III, we add nitric acid to the clear solution, to provide a high concentration of hydronium ions (its nitrate ions are again spectators). The hydronium ion, as an acid, reacts with the base, ammonia:

$$(3) \ NH_3 + H_3O^+ \longrightarrow NH_4{}^+ + H_2O.$$

And this reaction goes far to the right, for practical purposes removing ammonia molecules from the solution. This, of course, permits reaction (2) to reverse, setting free the blanketed silver ions, which can again be grabbed by the chloride ions that are still in the solution, to repeat the precipitation of reaction (1). See Figure 13-4.

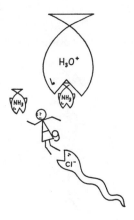

**Figure 13-4**

All of these reactions can be summed up in the set of equations below, where the labels on the arrows show the reactions that happen in the three steps of the test.

$$Ag^+ + Cl^- \xrightarrow[\xleftarrow{\text{Step ll}}]{\text{Step I}} AgCl \ (s)$$

$$+$$

$$2H_3O^+ + 2NH_3 \xrightarrow[\xrightarrow{\text{Step lll}}]{\text{Step lll}} 2NH_4^+ + 2H_2O$$

Step II ↓↑ Step III

$$Ag(NH_3)_2{}^+$$

These equations are chemically correct, but I have an idea that the goings-on are most easily understood if they're presented in a less formal way; hence the sketches accompanying the text where the little lady represents the silver ion. Neither the ammonia molecules nor the chloride ions have very sharp teeth. And, of course, when a hydronium has swallowed the ammonia, this can no longer maintain its hold on silver, who is again abandoned to the mercies of the chloride.

*What Happens with the Other Ions?*

The iodide and bromide ions, on the other hand, *do* have sharp teeth, as far as silver is concerned. Once they have grabbed the silver ion, it stays grabbed, because their solubility products are about $8 \times 10^{-17}$ and $5 \times 10^{-13}$, respectively. See Figure 13-5. Therefore, when they are in equilibrium with their saturated solutions, there are not enough silver ions in the solution to react appreciably with ammonia. These two are therefore eliminated at Step II, and we needn't worry further about them.

**Figure 13-5**

As for the other precipitates, the ones that did dissolve in ammonia in Step II, but didn't reappear (permanently) at Step III—we'll handle these by looking at silver phosphate. No new treatment is needed for Step II: the same ammonia complex forms there as when the original precipitate was the chloride.

In Step III, if the ammonia is precisely neutralized by the nitric acid, the precipitate will reappear, as the insoluble phosphate is reformed. But now a new factor enters the picture: *excess* acid will attack the phosphate ion itself.

The phosphate ion is a strong base (or, to state it in another way, the acids of phosphate are weak, especially $HPO_4^=$ and $H_2PO_4^-$). As excess nitric acid is added, the hydronium ion will react with the phosphate (Figure 13-6). (See Chapter 9.):

$$(4)PO_4^\equiv + H_3O^+ \longrightarrow HPO_4^= + H_2O.$$

With more acid, some $H_2PO_4^-$ ion will be formed as well. All this means that, although the silver ion is now free to precipitate, having been released from the ammonia, the phosphate ion is not:

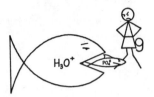

**Figure 13-6**

it has been "swallowed up" by the hydronium. Consequently, the precipitate either will redissolve, or (if the acid was added rather quickly) won't appear at all.

**Figure 13-7. A pictorial summary of the processes that may occur in the course of a chloride test.**

All the processes I've mentioned are shown in Figure 13-7, where the phosphate ion is the shark-like creature.

But what about all the other negative ions that formed insoluble silver salts? In our demonstration, we tried a couple of these—the carbonate and the chromate. The result (and the explanation) was exactly the same as with phosphate: hydronium attacked them and converted them to their respective acids. Usually the carbonate doesn't even reprecipitate at exact neutralization, and as excess acid is added, it fizzes a little, giving carbonic acid, which decomposes giving carbon dioxide gas.

But surely not *all* negative ions are strong bases! True: the chloride ion wasn't—HCl is a strong acid—but this is just the one we did *not* want to remove in Step III. How about sulfate, then? Fine: sulfuric acid, and the $HSO_4^-$ ion, are indeed strong acids. But neither of these makes a precipitate with silver ion in the first place. So since Step I never happened, we have no difficulties at Step III.

Let's go down the list of strong acids. Hydriodic and hydrobromic are strong, but we eliminated them at Step II, where silver bromide and iodide didn't dissolve. How about such strong acids as chloric, perchloric, trichloroacetic, etc.? It just happens that none of these, nor any other strong acid that I know of, has an ion that precipitates with silver.

To summarize, then, if the negative ion of a *weak* acid makes a precipitate in Step I, and this dissolves in Step II, the ion is removed in Step III. But if the ion of a *strong* acid makes that precipitate in Step I, ammonia (Step II) won't dissolve the precipitate—unless it is a chloride. The equations below show more formally than my fishy picture (Figure 13-7) the series of reactions that might occur in a "false positive" chloride test—one that gives positive results in Steps I and II, but breaks down at Step III. The symbol X stands for the element or radical of an acid HX.

$$Ag^+ + X^- \xrightarrow[\quad\text{Step II}\quad]{\text{Step I}} Ag X (s)$$

$$2H_2O + 2NH_4^+ \xleftarrow{\quad} 2H_3O^+ + 2NH_3$$

Step III

Step III ↑↓  Step II          Step III

$$Ag(NH_3)_2^+ \qquad HX^- + H_2O$$

$H_3O^+$

### The Sulfate Test

Some of the principles we've just described provide the rationale for another test that is familiar to every chemistry student. If barium chloride is added to an unknown solution, and a precipitate forms that will not dissolve in hydrochloric acid, a sulfate is present.

The explanation is relatively simple. The barium ion, like silver, forms precipitates, most of them white, with a good many negative ions. Besides sulfate, there are also carbonate, sulfite, and oxalate, for example. But all of these except the sulfate form weak acids in the presence of hydronium ions. The negative ions are thus tied up in acid molecules, permitting the precipitates to dissolve, just as the corresponding silver salts did. The only one that remains as a precipitate is the sulfate, an ion that has no appreciable tendency to pick up a proton.

The standard procedure is to use hydrochloric acid as the source of hydronium ions for this test, instead of nitric. I have wondered why, and can only suppose that the people who first devised the test realized that chloride ion was already present in barium chloride (the conveniently available source of barium ion), so no new spectators were being added. There is one more factor, though. I said that the sulfite ion gives a white precipitate with barium chloride. This ion is easily oxidized to sulfate—and nitric acid is a good oxidizing agent. There would therefore be a chance that the nitrate ion would not be the harmless spectator here that it was in the chloride test.

These last facts can be used to set up an intriguing puzzle demonstration. Start with a sodium sulfite solution, and add barium chloride to it. A white precipitate will form. If hydrochloric acid is now added, the precipitate dissolves, showing that it was not the sulfate*. But if now a little potassium permanganate is added to the solution, two interesting things happen: the purple color of the permanganate disappears instantly and a heavy white precipitate forms, which will not dissolve if more acid is added.

## The Sulfides

In years past, while most chem labs had an unpleasant smell, the analytical lab was the most nearly unbearable. The "rotten egg" gas, hydrogen sulfide, pervaded the place. Nowadays, two factors have tended to mitigate the problem. First, once it was realized that hydrogen sulfide is not only unpleasant but quite poisonous,

---

*Almost all samples of sulfites are contaminated by sulfate; therefore, the solution rarely clears completely at this point. Some slight turbidity almost always remains.

ventilation was greatly improved. And, second, many new methods have been devised for analysis, bypassing the gas. But it still remains useful, and its chemistry illustrates especially well some of the things we've been dealing with.

A great many sulfides are insoluble in water. The word "insoluble," of course, is not very precise. Iron sulfide is insoluble with a solubility product of about $10^{-19}$, manganous sulfide has a $K_{sp}$ of about $10^{-13}$, and the $K_{sp}$ of platinum sulfide is said to be $10^{-72}$. If taken seriously, that would mean that a Pacific Ocean-full of water would be needed to hold a couple of ions in a saturated solution of platinum sulfide!

Many of the sulfides have characteristic colors. Cadmium sulfide is a pigment painters call "cadmium yellow." Antimony sulfide is dark red, zinc sulfide pure white. These colors alone can sometimes serve to identify a specific metallic ion when hydrogen sulfide is bubbled into an unknown solution.

But the main reason hydrogen sulfide has been so useful for analysis is that the concentration of the sulfide ion in its solutions can be juggled over a very wide range. By doing this, taking the solubility products into account, it may be possible to precipitate one sulfide after another from a solution, each essentially pure.

A simple experiment shows this. Make up a solution that is around 0.1M in zinc ions, in copper (II) ions, and in manganese (II) ions. Use sulfates, nitrates, chlorides—whatever bottles are handy. Take 5-10 ml of this mixed solution in a test tube, add a milliliter or two of dilute nitric acid to it, and bubble hydrogen sulfide gas through for a few minutes. And if my instructions seem rather casual, that's just the point: the amounts of the materials don't matter very much.

A heavy black precipitate of copper sulfide will form, and if the $H_2S$ is bubbled through for two or three minutes, no appreciable amount of copper ions will remain in the solution. The mixture is filtered, and the filtrate saved. (In careful work the precipitate is washed with water, and the washings are added to the filtrate.)

Now a gram or two of solid sodium acetate is added to the clear filtrate. Immediately, white zinc sulfide appears. More $H_2S$ is bubbled through until all the zinc sulfide has precipitated. This is again filtered off (the mixture may be heated to boiling for a moment before filtering if desired; this coagulates the precipitate: makes its crystals bigger, so they filter better), and the filtrate is again saved.

To the final filtrate is added some concentrated ammonia. Flesh-colored manganous sulfide should appear at this point, and can all be precipitated with more $H_2S$.

## Juggling the Ions

The separation depends on changes of pH of the solution, with consequent changes in sulfide ion concentration. The first solution was strongly acidic, with a pH of 0-1. Then we added the acetate ion, a weak base, which brought the pH up *toward* neutrality, by producing a buffer (acetic acid-acetate: cf Chapter 11). And finally the added ammonia moved the pH to, or a little above, neutrality, with an ammonium-acetate buffer. What do these changes do?

Hydrogen sulfide is of course a diprotic acid, ionizing in water in two stages:

$$H_2S + H_2O \longrightarrow HS^- + H_3O^+ \qquad K_I = 10^{-7}$$

$$HS^- + H_2O \longrightarrow S^= + H_3O^+ \qquad K_{II} = 10^{-15}.$$

The size of the second of these constants indicates that the $HS^-$ ion is a very weak acid indeed—weaker than water itself.

In a saturated solution of the gas, it may be assumed that the concentration of dissolved $H_2S$ molecules is about 0.1M. By simple algebraic manipulation of the equilibrium expressions with this molecular concentration*, it can be shown that the sulfide ion

---

*The derivation of the expression relating hydronium ion to sulfide ion in saturated $H_2S$ solution follows:

The first ionization gives: $\dfrac{[HS^-][H_3O^+]}{[H_2S]} = 10^{-7}$.

The second gives: $\dfrac{[S^=][H_3O^+]}{[HS^-]} = 10^{-15}$.

If these two equations are multiplied together, the term $[HS^-]$ cancels out, since we are speaking of the actual concentration of this ion in a single solution. Thus we get:

$$\frac{[S^=][H_3O^+]^2}{[H_2S]} = 10^{-22}.$$

If we now assume that in a saturated solution of $H_2S$ gas the concentration of the molecules of the gas—$[H_2S]$—is about 0.1M, the final result is $[S^=][H_3O^+]^2 = 10^{-23}$.

I must add that there is no general agreement on the exact size of the second ionization constant for $H_2S$: three books that I picked up in succession gave three different "constants." I have therefore picked a value that gives calculated results that are consistent with experiment. I have taken the same liberty with the solubility products I've used, about which, again, there is surprisingly little agreement!

concentration is related to the hydronium ion by the equation:
$$[H_3O^+]^2 x[S^=] = 10^{-23}.$$

That formula is a rather sterile way of expressing a fact that is more easily grasped by looking at the chemical equations. These show that when $H_2S$ ionizes, a very small amount of sulfide ion is produced at the end of two reactions, in both of which hydronium is also produced. The hydronium produced in the first reaction tends to inhibit the second one. If *excess* hydronium is provided by adding another acid, the reactions are pushed backward, and even less sulfide is produced. But if the hydronium concentration is *decreased*, by adding a base, then the forward reactions will be encouraged, and more sulfide ion will appear in the solution.

Returning now to the quantitative; in 1N acid, where the $[H_3O^+]$ is about 1, the $[S^=]$ is $10^{-23}$; whereas, in a solution buffered at pH 5, the $[S^=]$ is $10^{-13}$, and at pH 10 it will rise to $10^{-3}$. Each of these changes is a 10-billion-fold increase in sulfide ion concentration! The full relationship can be plotted on a graph where the sulfide ion concentration (expressed in powers of 10) is plotted against the pH. See the slanting line in Figure 13-8.

Now put these facts together with the solubility products of the metallic sulfides we are considering. Approximate values of these are:

$$CuS\ K_{sp} = 10^{-36}$$
$$ZnS\ K_{sp} = 10^{-22}$$
$$MnS\ K_{sp} = 10^{-13}.$$

In describing the demonstration, I suggested that the initial concentrations of the ions be made roughly 0.1M. If the nitric acid added at the beginning had made the solution 1N in acid, then the $[S^=]$, as mentioned above, would have been $10^{-23}$, and the ion-product for each of the possible metallic sulfides (i.e., the product of the concentrations of sulfide ion and metallic ion actually present) would have been $0.1 \times 10^{-23}$, or $10^{-24}$. This figure is a trillion times as large as the solubility product for copper sulfide, but much smaller than the $K_{sp}$ values for either zinc or manganese sulfide. Therefore, all the copper ion was precipitated from the solution, while both zinc and manganese ions remained soluble and passed into the filtrate. The extra hydrogen sulfide bubbled through the solution of course replaced the $H_2S$ molecules that were removed as the copper sulfide precipitated.

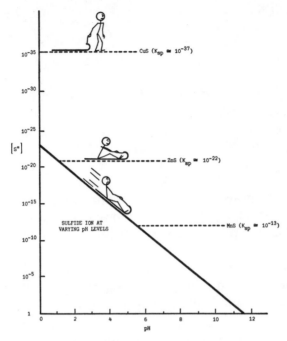

Figure 13-8. Precipitation of sulfides from 0.1M
solutions of metallic ions.

When sodium acetate was now added to the clear filtrate, the acetate ion as a base reacted with the hydronium of the nitric acid, forming acetic acid. This, with the excess acetate ion, could be expected to give a pH in the range 2-5 (the buffer range for acetate-acetic acid). If we suppose that the pH was 3, the $[S^=]$ would be $10^{-17}$ (see Figure 13-8), and the ion-product for each of the metallic sulfides now present would be of the order of $10^{-18}$. This is now much bigger than the $K_{sp}$ of zinc sulfide, but less than that of MnS, so the former precipitated.

Finally, with the addition of ammonia, the pH again was raised, this time to a value of 6 or more, and the increased sulfide ion was easily enough to make the manganese sulfide precipitate.

Figure 13-8 can be adapted to fit any set of $K_{sp}$ values, and indicate about where precipitates should form. However, these values (and therefore my toboggan-slide) have to be used with caution. Factors such as temperature, total concentration of ions in solution, degree of supersaturation, etc., all can affect experimental results when $K_{sp}$ values are close together.

*Further Identification*

Further identification of the three sulfides we dealt with here can be done in a number of ways. Here is a sample.

*Copper:* Copper sulfide was precipitated from a strongly acid solution, and its solubility product is so small that extraordinary measures have to be taken to get the copper ion back into solution. This can only be done by destroying the sulfide ion. If the black precipitate is scraped off the filter paper, and then boiled with dilute nitric acid, the sulfide is oxidized to sulfate. Copper sulfate is soluble, so the characteristic blue color of the copper ion reappears. For further identification, an excess of ammonia can be added. After the nitric acid has been neutralized, the beautiful indigo-blue color of the cuprammonium ion, $Cu(NH_3)_4^{+2}$ appears.

*Zinc:* The zinc sulfide precipitate can easily be dissolved by merely adding some hydrochloric acid. The hydronium of this combines with the sulfide in the type of reversal we saw with silver phosphate:

$$Zn^{++} \ + \ S^{=} \ \rightleftharpoons \ ZnS(s).$$

$$+$$

$$H_3O^+$$
$$\downarrow$$

$$HS^- + H_2O$$

The zinc ion can then be identified, if this is thought necessary, by again neutralizing, and adding hydrogen sulfide again. I believe that the only insoluble *white* sulfide is that of zinc.

*Manganese:* Finally, the manganese sulfide precipitate will dissolve in a solution that is only faintly acidic. A drop or two of acid is all that is needed. Identification of the manganous ion can be done by adding a vigorous oxidizing agent, like sodium bismuthate or periodate, which yields the familiar purple color of the permanganate ion.

*Other Metallic Sulfides*

Although I have given details for the sulfide precipitation of only three metallic ions, many others can be handled. Each of the three I chose is typical of a group: those which will precipitate no matter how strong the acid, those which will appear only in weakly acidic solutions, and those which will not appear unless the solution is neutral or basic.

The identification of the very insoluble sulfides presents some interesting problems, since the metallic ions must first be returned to solution before they can be identified. Some of the sulfides, like that of copper, will dissolve if the sulfide ion is destroyed with warm nitric acid, a powerful oxidizing agent. But mercury (II) sulfide and some others are too insoluble even for this. In such cases, aqua regia works: a mixture of concentrated hydrochloric and nitric acids. It is probable that aqua regia acts by forming chloride-ion complexes with the metallic ions, at the same time as the nitric acid attacks the sulfide ion. Under this two-pronged attack, even an excessively insoluble sulfide can be torn apart.

In the case of some of the sulfides, another type of complex is effective. For example, arsenic and antimony sulfides will dissolve in a solution containing excess sulfide ions, because the soluble $AsS_4^{-3}$ and $SbS_4^{-3}$ ions are formed. This process is a strange reversal of the ion-juggling we have been indulging in. Up to now, more sulfide has meant more precipitate, but here, excess sulfide ions actually dissolve the precipitate.

## Titration of the Chloride Ion

At the start of this chapter, I said that for the quantitative analysis of a solution, precipitates can be filtered out and measured. The oldest method of analysis for the chloride ion is to precipitate the silver salt, dry it, and weigh it. But this procedure is tedious. The precipitate must be collected in specially prepared and weighed filters. It must be painstakingly washed, to remove any other weighable material. And finally, with the filter, it must be dried and weighed again. Anyone who has been through a course in "quant" (quantitative analysis) has been through these steps.

But all this weighing and drying can be bypassed with the use of a quick titration method. The solution to be analyzed is simply titrated with silver nitrate solution of known concentration, using a few drops of potassium chromate as the "indicator." The standard silver nitrate solution is added from a burette, slowly, with constant stirring. At first, the white precipitate of silver chloride is the only thing seen. As the endpoint is approached, some bright maroon color appears in the neighborhood of each added drop, which fades upon stirring. Finally, as the first excess drop of silver nitrate is added, the brilliant color of the silver chromate appears throughout the solution, and remains—as characteristic a color change as that seen with acid-base indicators.

The method depends on two facts: First, obviously, on the bright color of the chromate. But second, on the fact that the solubility of silver chloride is much less than that of the chromate*. In a saturated solution of pure silver chloride (i.e., a solution in contact with the precipitate), the $[Ag^+]$ is about $10^{-5}$. For pure silver chromate, the figure is about $10^{-4}$.

As each drop of the silver nitrate solution mixes with the chloride solution, the solubility product of silver chloride is exceeded, and the white silver chloride is formed, removing the added silver. But when the chloride concentration drops below $10^{-5}$, the silver ion can rise *above* $10^{-5}$, quickly reaching the concentration needed to form the brightly colored silver chromate.

As a practical lab process, a chloride determination can be done in this way in a few minutes. This compares with a day or so that is needed for the gravimetric procedure (counting drying time, as well as the filtration and weighing). So it's not surprising that technicians prefer the titration method. There are a couple of precautions to be observed, however. The most important is to be sure that chloride ion is the only one present that can form a very insoluble precipitate with silver. In addition to this, if the solution to be measured is acidic, the acid must be carefully neutralized before the titration, otherwise the chromate ion would be tied up by the hydronium (see the first part of this chapter) and not be available for its color formation with the silver ion. One routine use of this titration is to measure the "salinity" of naturally occurring water.

## Conclusion

In this chapter, we have been able to touch on some of the problems of the analytical chemist. His methods have been

---

*There's no point in working with the technicalities of $K_{sp}$ values here. Since the constant for chromate involves the square of the silver ion concentration, there is no simple way to visualize the comparison of the figures.

developed over a period of many centuries—some of them first used simply because "they worked." A great deal of ingenuity has gone into the refining, and the understanding, of these important reactions.

# 14

## Dynamic Equilibrium Explains
## Diffusion, Dialysis, and Osmosis

Diffusion sometimes seems like a rather vague idea. Molecules move around, sometimes at high speeds, mixing with each other. But what can diffusion *do*? One of the easiest ways to see diffusion "doing" something is to make a "Hydrogen Fountain." This is shown in Figure 14-1. A is a cup made of unglazed porcelain and B is a beaker big enough to contain the cup with room to spare. The flask at the bottom contains water. The beaker B is filled with hydrogen gas, and lowered over the cup A, whereupon water spurts for a few seconds from the spout in the lower flask.

Notice that this is not a simple pressure effect. The total pressure is the same—1 atmosphere—inside and outside the cup when the beaker is lowered over it.

Before the beaker is placed over the cup, molecules of air (oxygen and nitrogen) are constantly going in and out through the tiny pores of the cup. These molecules have an average kinetic energy that depends on the temperature of the room. The hydrogen molecules in the beaker have average kinetic energy which is the same as that of the air. But hydrogen molecules are

188

Figure 14-1

roughly 1/15 as heavy as the molecules in air. Hence, they must be moving roughly four times as fast, if their kinetic energy is to be the same (KE = $\frac{1}{2}MV^2$). When the beaker is lowered over the cup, surrounding it with hydrogen instead of air, the hydrogen molecules can move *into* the cup much faster than the air molecules can move out. So for a short time the pressure inside the cup rises.

But air is still diffusing out, though more slowly than hydrogen is diffusing in, and soon the concentrations of all the gases become even and the "fountain" stops. If the beaker is now removed, air will bubble into the lower flask through the spout tube, showing a pressure drop as hydrogen leaves the cup.

In Chapter 4 we were exposed to the idea of entropy, which describes, and can be used to measure, the degree of disorder of any system. We saw there that increase in entropy can actually be the driving force causing something to happen. To relate our fountain to the concept of disorder, notice that when the beaker was first lowered over the cup we had a relatively ordered situation, with mostly hydrogen outside and mostly air inside. The diffusion that built up the pressure, while explained locally in terms of the speeds of the molecules, was in the broadest sense the result of the tendency of any ordered system to give way to a disordered one where possible.

The principle of the hydrogen fountain is easily extended to liquids, for example to situations where solutions of different concentration are in contact with one another. If you take a tall cylinder of water and carefully sink some potassium permanganate crystals to the bottom of it, they will dissolve there, making a purple layer at the bottom of the cylinder. If the cylinder is now left undisturbed for a long period (weeks may be needed), the purple color will gradually creep up. The disorderly thermal motion of the potassium and permanganate ions, and of water molecules, slowly overcomes the original orderly separation, and "lifts" the denser liquid from the bottom of the cylinder.

## Dialysis or Osmosis

The words dialysis and osmosis are almost synonymous, but *osmosis* is always used when pressure effects are being considered. Dialytic processes are of vital importance in all biological systems. Unfortunately, these processes are given rather slighting treatment in most elementary chemistry books.

Many types of membranes will permit slow passage of small molecules and ions in solution. This phenomenon is called dialysis, and is closely related to the gaseous diffusion that produced the Hydrogen Fountain. It can occur through almost any biological membrane—the wall surrounding a single cell or the meters-long tube that is the intestinal tract. In fact, much early laboratory work on dialysis was done with animal membranes. Nowadays, collodion, cellophane, and more modern synthetics are routinely used.

If a solution of potassium permanganate (deep purple) is poured into a cellophane or collodion bag*, and this is then lowered into a container of pure water, the color of the permanganate immediately begins to "leak" through the membrane. See Figure 14-2. Yet gentle pressure can show that there is in fact no true leakage: there are no holes in the membrane, even of a size detectable with a microscope. The effect is less obvious with colorless solutes, but any salt will do the same thing. It seems that salts can diffuse,

---

*The process for making a membrane sac is simple. Coat the walls of a large test tube with collodion solution by pouring some of the solution into the test tube, and then back into its original container, rotating the test tube constantly as this is done. Then suspend the tube, mouth down (over a paper towel to catch the drip) for 10 to 15 minutes, until most of the ether has evaporated from the collodion coating. Now fill the tube with water, and let it soak for at least 10 minutes (it can be left for days if desired: the membrane should not be allowed to dry out). Finally pour out the water, cut the film around the mouth of the test tube, and gently disengage the membrane sac. It is wise to test the sac for holes by filling it with water and blowing gently.

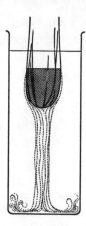

Figure 14-2

ion by ion, through the material of the membrane. Small molecules, like alcohol or sugar, can do the same thing.

But there are many types of materials that will not dialyze—will not diffuse in this way through a membrane. These include very large molecules such as those of proteins (egg albumin, gelatin) or polysaccharides (glycogen, starch), as well as the whole range of "colloids." This last is a vaguely defined, but exceedingly important, state of matter lying between true solutions and suspensions which will settle out. Rough definitions of colloids sometimes speak of particles with masses upwards of 10,000 amu.

Two convenient colloids for laboratory demonstrations are those of arsenious sulfide, $As_2S_3$, and ferric hydroxide, $Fe(OH)_3$. To make the first, a saturated solution of arsenic trioxide, $As_2O_3$, is prepared by boiling a little of the powder in water, cooling, filtering, and then bubbling hydrogen sulfide gas through the filtrate for a short time. A clear yellow liquid results, not easily distinguishable from a true solution. The iron hydroxide colloid is even easier to make: about 1 ml of 1M ferric chloride is added to half a liter or so of water, and the mixture brought to a boil. Hydrolysis of the hydrated ferric ion (see Chapter 9, page 124), produces a clear, deep red liquid. This is again not easy to recognize as anything other than a true solution. Yet if either of these two liquids is put in a dialysis sac (one that will easily pass potassium permanganate), none of the color will diffuse out. In each case, the particles that produce the color are made up of many thousands of atoms each. They are simply too big to pass through the pores of the membrane, which are of common molecular dimensions.

Colloidal particles do not settle out of solution for a number of reasons. The primary one is their bombardment by fast-moving molecules of water (this is called Brownian movement). Some colloidal particles carry a dense layer of water, and many of them carry electrical charges, all the particles of a given colloid having the same charge. The arsenic sulfide particles are negatively charged, the iron hydroxide ones are positive.

I mentioned the fact that some biological molecules, like proteins, are of colloidal dimensions. Hence, a standard method for purification of these in the laboratory is to dissolve them in water, put the solution into a dialysis sac, and surround this with water. Small molecules, including sugars and salts, will diffuse through the membrane as the system moves toward greatest disorder. If the external water is constantly renewed, all small molecules can be removed from the colloidal suspension. We'll return to this effect later in connection with the functions of the lung, the kidney, and the single cell.

## Osmotic Pressure

I said that osmotic effects are closely related to the Hydrogen Fountain. But there was a very clear pressure effect in the Fountain that I haven't yet mentioned in connection with osmosis.

When I described the diffusion of potassium permanganate through a membrane, I may have spoken as though its ions were the only small particles present. But there is also *water* itself on both sides of the membrane. If potassium and permanganate ions, with masses of 39 and 119 amu, could pass through easily, then certainly so could water, with a molecular mass of only 18. With potassium permanganate, a pressure effect would not be easy to demonstrate, because all the particles pass so easily through the membrane. But if we substitute a solution of cane sugar (molecular weight 342) for the permanganate, then we begin to see something comparable to the Fountain.

In a tightly closed membrane, with a glass tube coming out of the top, a sugar solution may cause the water level in the tube to rise several feet above the water outside. As with the Hydrogen Fountain, this is a transient phenomenon. We'll examine it in a little detail.

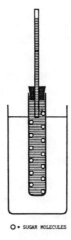

O = SUGAR MOLECULES

**Figure 14-3**

When the sac with the sugar solution is first put into pure water, there is diffusion of molecules (all the molecules present) through the membrane in both directions. See Figure 14-3. The temperature (and hence the average kinetic energy) of all the molecules is the same, so the small molecules of water are moving much more quickly than the sugar molecules (cf. the first page of this chapter). Water alone will be bombarding the outside of the membrane, while both types of molecules will be bombarding the inside. Since the concentration of water on the outside is greater than on the inside, the total flow of water into the sac must exceed its flow out. (Notice the rather unusual point of view here; we consider the concentration of *solvent,* rather than of solute. We did this once before, when discussing Raoult's Law, in Chapter 6.) This inequality of water flow is partly balanced by the flow of sugar molecules out of the sac, but as we have seen, this flow is slower than that of water, since the sugar molecules are moving more slowly. Consequently, net pressure begins to build up inside the sac.

Another way of looking at the same situation is to consider that the sugar molecules "get in the way" of water molecules which would otherwise escape from the sac. We used this kind of treatment when we were considering the effects of solutes on the vapor pressure, freezing, and boiling of liquids (Chapter 6).

Whichever of these models appeals to you—and they are both useful—the process is another example of the tendency toward increased entropy. We begin with the relatively ordered state that has all the sugar on one side of the membrane. This slowly gives way to a situation of equal disorder on both sides, as the sugar solution inside is diluted, and sugar concentration builds up outside. The build-up of pressure is caused by the unequal speeds of these two processes, exactly as happened with the Hydrogen Fountain.

But now let's move on to the situation where the solute inside the sac is of colloidal dimensions: where the suspended material *cannot* get through the walls. The pressure inside will build up in just the same way, but now there will be no long-run leakage of the material that caused the imbalance. Nature still inexorably demands her increase in disorder. The colloidal suspension inside the sac *must* be diluted, and so there is a build-up of osmotic pressure. This colloid-caused pressure is often called "oncotic" pressure by physiologists. If the sac is not sealed, water will continue to flow in, dilute the colloid, and eventually flush it out. But if the sac is sealed, pressure will build up, sometimes very high pressure. If the membrane does not burst, balance will finally be achieved in a new way. As the pressure mounts inside, water is forced out more rapidly—no longer by simple diffusion, but by actual flow through the pores of the membrane: it is literally "squirted" out, past the interference of the colloidal particles. In this way, equilibrium of flow will be achieved—inward on account of the higher *concentration* of the water outside, outward on account of the higher *pressure* which offsets the lower concentration.

## Osmotic Pressure Measurement

Under these conditions of equilibrium, calculation of osmotic pressure is possible. The reasoning seems paradoxical, and the results surprisingly simple.

When pressure inside a sac has been allowed to build up until equilibrium is attained, water is flowing at equal rates into and out of the sac. If this is the case, then the *rate of flow of the water* cannot be causing the pressure! What then does cause it? The chemist van't Hoff came up with the answer in 1886. Under the equilibrium conditions, both the inside and the outside of the sac are being subjected to equally effective bombardment by the water molecules (the lower concentration of water inside being offset by its higher pressure). But the inside of the sac is also being bombarded by the colloidal particles. These particles may be considered, at least in dilute solution, to be as free-moving as though they were in the gaseous state. Hence, their pressure on the membrane should be as great as the same number of molecules of gas confined in the same volume. Thus, in the equilibrium situation, we suddenly look away from the *water* flow, which seemed to cause the pressure, and look instead at the colloidal particles and their concentration.

If we had 1 mole of any substance, *as an ideal gas,* forced into a 1-liter container at 0°C, its pressure would be 22.4 atm. Then 1 mole of a colloid, contained in 1 liter of solution within a membrane, should exert the same pressure. Experiment has borne out this hypothesis, for dilute solutions. In practice all colloidal solutions *are* very dilute, in terms of moles per liter. And the formula for the General Gas Law, $PV = nRT$, applies remarkably well to these solutions.

The formula for osmotic pressure that I've just given can be derived in a totally different way, and usually is, in physical chemistry texts. The model that I have presented here is however perfectly clear, and readily comprehensible. It emphasizes again the extraordinary versatility of Avogadro's original hypothesis— that equal numbers of molecules have equal effects under a wide variety of conditions, in spite of tremendous variation in the size of the molecules. In the case of the colloidal polysaccharide like glycogen, for example, the individual particles may weigh as much as 13 *million* amu, yet the law of $PV = nRT$ still applies, where n is the number of moles of glycogen dissolved in volume V of water.

By the use of van't Hoff's Law, accurate measurements can be made of the particle mass for substances of very high molecular weight, like proteins, enzymes, hemoglobin, etc.

*Particle Weight Measurement*

Suppose we want to find the molecular weight of a water-soluble substance. The figure is suspected to be about 10,000 amu.

We saw in Chapter 6 that one routine method for molecular weight measurement is to determine the freezing point depression for a solution of known concentration. So we make up a 10% solution of our unknown: 100 g added to a kilogram of water. This will be a 0.01 molal solution, and its predicted freezing point will be in the neighborhood of -0.019°C (i.e., a depression of 0.01 times the freezing point constant for water: cf Chapter 6). It is very difficult to measure with any accuracy a freezing point depression as small as this: an error of a thousandth of a degree will introduce a 5% error!

But now suppose we measure the colloidal osmotic pressure of this same solution. The gas laws predict that 0.01 mole of gas in a 1-liter container at 0°C would have a pressure of 0.01 times 22.4 atm, or 0.224 atm. If we were to measure this pressure with a column of water, like the one shown in Figure 14-3, the column would be better than 5 feet high, and the pressure could therefore be measured to any desired degree of accuracy. Refinements of this basic method have made possible the measurement of colloidal particle sizes as great as several million amu, like the glycogen mentioned above. A solution of this sort will give *no* measurable freezing point depression.

The word *colligative* must be introduced here. It is a word not often used—but often enough so its meaning should be known. Four of the properties of solutions that we have discussed in the course of this book—vapor pressure, boiling point elevation, freezing point depression, and osmotic pressure—are all directly related to the molal concentration of the solute, and are all caused by the same action of the solute. In each case the solute interferes, particle by particle, with escape of solvent molecules. Since the properties are tied together in this way, they are called the *colligative* properties, which comes from a Latin word that means literally "tied together."

*Biological Applications*

So far I have dealt with laboratory systems, which work well to illustrate the basic principles. Biological applications of these principles are so widespread that I can take only a sampling.

I still remember how, as a boy, I looked at a drop of blood under the microscope my mother had used in college. The red blood cells were crowded together too closely to be seen clearly, so I added a drop of water to dilute the blood. I had done this with fine dust and other things. The dilution was worse than useless. The red cells simply disappeared. Finally my mother suggested that I try salt water, which kept the cells beautifully intact. Each cell is surrounded by a membrane, and inside is a solution of all kinds of molecules and ions, colloidal and otherwise. In whole blood, the cell is bathed in a solution that is *isotonic* with it (in osmotic equilibrium). But when these little sacs are put in pure water, then fluid pours into the sac faster than even the smaller molecules can escape. Pressure builds up, and the cell simply explodes. This is why, in a hospital, a patient in need of fluids is given "saline solution" intravenously, rather than water.

The reverse process happens when fresh flowers or vegetables are cut and allowed to stand where moisture can evaporate. Here water vapor can diffuse out (this is called transpiration) and none gets in. The plants wilt. But a short immersion in water will often perk them up, as osmosis causes flow of water into the cells.

There is a complex set of dialysis processes (at first sight involving no pressure effects) during the circulation of blood in an animal. We'll look at it rather superficially. Begin with blood that has just left the lungs, charged with oxygen (in solution, and in loose combination with hemoglobin: the latter increases the carrying capacity of the blood, but does not alter the course of the processes we're considering). This oxygen-rich blood goes to the heart, and is then pumped past the stomach and intestines, where it picks up small molecules—products of digestion that have dialyzed out through the intestinal membrane and in through the vein walls. The blood carries this load to the liver, where there is exchange (again by dialysis) of small molecules as the liver carries on its "factory" function.

Moving further, the blood reaches fine capillaries throughout the body, where it is close to the individual cells. In these cells metabolism has been going on, so there is a lack of oxygen and an excess of carbon dioxide; also a lack of nutrients, and an excess of waste materials such as urea. For each of these species, nature insists on a maximum of disorder; i.e., there should not be higher concentration of any one of them on either side of the cell wall. Consequently, dialysis occurs, with a steady flow of oxygen into, and carbon dioxide out of, the cell; also of nutrients in and waste products out. The blood now moves out of these capillaries into new veins and presently passes the kidneys.

Kidney function is now known to be far more than a simple "passive" osmotic process, but normal dialysis does occur for many substances. Urea, for one, simply diffuses through kidney membranes, to equalize its concentration within and without the blood vessels. Some salts do the same thing. Thus these waste materials of metabolism are swept away and excreted in the urine. Why sugar, and many specific ions, are *not* lost in this way is still far from clear. There is definitely an active "reverse osmosis" that occurs in the lower glomeruli, which returns these materials to the blood against the osmotic gradient, with the expenditure of energy.

Finally the blood is pumped again to the lungs (I have greatly oversimplified the circulatory network), where again there are concentration differences between oxygen and carbon dioxide on the two sides of the lung membranes. Carbon dioxide diffuses out of the blood into the alveolus (the air pocket) at the same time as oxygen from the air diffuses into the blood. The function of hemoglobin is to provide still another equilibrium step. Oxygen can combine with hemoglobin in a loose chemical combination, very easily reversible. The quantity of oxygen present in true solution is actually not very large, but it is "backed up" by the oxyhemoglobin (very much as the hydronium ion in a buffer solution is backed up by the acid of the buffer). And there is even another equilibrium step involving gases: the carbon dioxide is readily and reversibly converted into bicarbonate ion. This reaction was discussed in Chapter 11, when we were dealing with the buffer systems of blood. But the reaction permits the blood to carry more carbon dioxide than would be present in true solution.

## Osmotic Pressure and Circulation

In dealing with circulating blood, I have spoken so far only in terms of diffusion, or dialysis. But there are important pressure effects. We'll cover one of these in some detail.

The individual cells of the body are in fact not directly bathed by the blood, but by the "interstitial fluid," which has no circulation comparable to that of blood itself. This fluid moves very effectively, however, from blood vessel to cell, carrying its life-giving nutrients, then moving back again with its waste cargo to the blood vessel. What causes the motion?

Blood is a very complex substance, containing a large amount of colloidal material. Since this material cannot leave the vessel it will tend to make water "leak into" the vessel osmotically, unless there is pressure inside to balance this tendency. As blood reaches a capillary from arteries and arterioles, there *is* such a pressure— simply the blood pressure, furnished by the heart-beat (plus hydrostatic pressure, in lower parts of the body). In a normally operating body, this pressure is great enough, not only to balance, but also to overcome the colloid osmotic pressure and force leakage of solution out of the capillary. It is this solution that carries small molecules in solution directly up to the cell wall, where the dialysis mentioned earlier can occur.

Meanwhile the blood, carrying its indiffusible colloids, passes through the capillary. See Figure 14-4. As it goes through the narrow passage, its pressure drops sharply (the venous pressure is barely adequate to return the blood to the heart). Thus at the venous end of the capillary, osmotic flow will occur normally: interstitial fluid will *return* through the capillary wall to dilute the colloids of the blood, and the blood will move on through the venous system, its exchange function effectively completed.

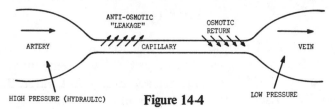

Figure 14-4

## Conclusion

In this chapter, I have barely skimmed the surface of the effects of diffusion and osmosis in biological systems. They can be studied in detail in areas as diverse as the flow of sap in trees, the metabolism of migratory fishes, the pigments in butterfly wings, and the effects of starvation on human beings. Much remains to be learned.

*A Sampling of Additional Demonstrations and Problems*

The build-up of osmotic pressure can be demonstrated with equipment even simpler than that shown in Figure 14-3 of this chapter. If a saturated sugar syrup is poured into a dialysis sac and this is hung over the edge of a cylinder of water (Figure 14-5), concentration waves will immediately be seen (like those in Figure 14-2) below the sac. But within half an hour, the level of liquid within the sac will have risen noticeably, and very soon the sugar solution will start dripping out of the sac (it's not hard to clean up this sugar solution on a table-top!).

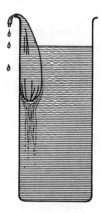

**Figure 14-5**

As for dialysis, I mentioned the fact that an iron hydroxide sol will not dialyze out of a sac. The original sol was made by adding a milliliter or so of 1M ferric chloride solution to hot water. But if 1 ml of the 1M ferric chloride solution is added to water inside a sac, *without* heating, there will be visible dialysis of the ferric chloride (technically, of the iron and chloride ions) through the sac. If the set-up is left for a day, most of the iron solution will have left the sac, but *some* will remain, and will not dialyze out. As the solution has become more dilute, the hydrolysis reaction begins to take on significance, even with cold water:

$$Fe(H_2O)_6^{+3} + 3H_2O \longrightarrow Fe(OH)_3(H_2O)_3 + 3H_3O^+$$

and the insoluble hydroxide forms particles of colloidal dimensions.

An interesting osmotic pressure calculation is the following. In hospitals, "normal saline solution" is used for intravenous injection of fluid. This is approximately 0.9% sodium chloride solution. *Supposing* a membrane could be designed that would permit sodium and chloride ions to pass through, but not water (this is in fact impossible). What would be the osmotic pressure at equilibrium inside a sac containing this saline solution? How high a column of water would it lift? The astonishing answer to the second question is that the water column would be something over *250 feet* high!

# 15

## Dynamic Equilibrium
## and Electrochemistry

Most high school students have at least heard of Galvani's "experiment with the frog legs." But a surprising number haven't, and the full significance of the experiment is rarely explained.

Between 1771 and about 1791, Luigi Galvani, an Italian physiologist and anatomist, studied a phenomenon that he had first observed by accident. When frog legs, hanging from copper hooks preparatory to being cooked, happened to swing against an iron railing, the legs twitched strongly. Galvani seems to have realized very early that the effect was related to electricity, and he was confirmed in this when the twitching was also seen in the neighborhood of an electric machine, or when lightning struck nearby. Most of the rest of Galvani's ideas were superseded, but he established firmly that for the original effect two different metals were needed, which touched one another at the same time as both touched the muscle. See Figure 15-1.

After many false leads had been explored, it was finally shown that a current of electricity was generated by the two metals in contact with the salt solution of the muscle. The muscle itself played a double role. It provided the salt solution, and the resulting current made it twitch, so it acted also as a galvanometer.

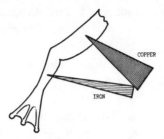

**Figure 15-1**

In 1800 Alessandro Volta made the first electric battery by stacking together a pile of copper and zinc discs, with pieces of moist paper (instead of frog muscle) strategically placed in the pile. See Figure 15-2. If the paper were moistened with acids or salts, and a tall enough pile used, considerable "voltage" could be produced (this electrical unit obviously being named for the inventor of the "pile").

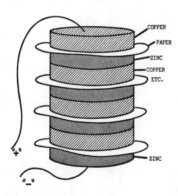

**Figure 15-2**

In the next 30 years, scientists excitedly investigated the effects of the electric current that was now available. This research culminated in the work of Michael Faraday, published in 1833. Faraday found that the quantity of electricity could be measured by its chemical effects, and that the same quantity always produced the same effects. Some of this work that seems obvious to us now must have been much less clear to those who did the experiments, since this was the period when Dalton's atomic theory was completely accepted, but Avogadro's ideas were not yet taken seriously. In any event, Faraday coined the term "equivalent," still used by chemists to describe the quantity of a chemical element that is produced by a given amount of current. He is also responsible for the words "ion," "electrolyte," and "electrode."

Faraday's findings in this connection were essentially the following. If enough electric current is passed through a solution of hydrochloric acid to free 1 gram of hydrogen gas at the cathode, then 35.5 g of chlorine gas will be formed at the anode. And if this same current is conducted through a series of solutions, say sulfuric acid, silver nitrate, copper chloride, etc., an "equivalent" of metal or gas is produced at each electrode in turn. See Figure 15-3. In terms of our present-day atomic weights (not necessarily the same ones that Dalton, and therefore Faraday, used), this equivalent is 8 g of oxygen, 1 mole of silver, 1/2 mole of copper, 1/3 mole of gold, etc. In each case, the equivalent is the number of moles of the element that would combine with 1 gram-atom of hydrogen or 1/2 gram-atom of oxygen. The amount of current that will produce this effect is now called, appropriately enough, a *faraday*, and is simply 1 mole of electrons (about $6 \times 10^{23}$ electrons). In electrical units, this is 96,500 coulombs (or ampere-seconds).

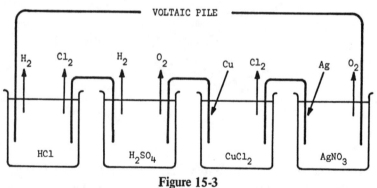

**Figure 15-3**

*Electrolysis: Molten Compounds, and the Basic Principles*

When teaching the details of electrolysis in modern terms, it is wise to start with the simplest possible system—the passage of current through melted sodium chloride with the use of inert electrodes (platinum is best). Even for students who are already familiar with the meaning of an electric current, some preliminary clarifications are needed.

When electricity passes through a metallic wire, it is in the form of moving electrons only. These exist as a "gas" capable of free motion throughout the solid matrix that is provided by the positive ions of the metal. Physics students have usually had drummed into them the idea that it is "only the negative that moves." When current is produced by a generator, moves through a circuit, and lights a light bulb, for example, the generator simply pushes the "electron gas" round and round the circuit.

But when a current produces electrical effects, this simple concept must be modified. In molten sodium chloride there are two kinds of charged particles, positive sodium ions and negative chloride ions. *Both* of these can carry current, and electrons take no part in the transfer of charge. Both ions can move freely, subject only to the restriction that in any tiny volume the total number of positive and negative charges must be equal. See Figure 15-4. This is an important restriction, and we'll return to it again and again. Whenever a sodium ion moves out of a region, for any reason, *either* another sodium ion must move in *or* a chloride ion must also move out.

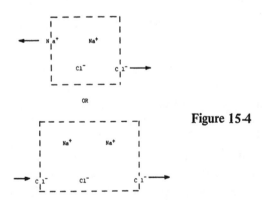

Figure 15-4

When current-carrying electrodes are put into the molten salt, the cathode (the "negative" pole in this case) will have an excess of electrons and the anode a deficiency of electrons. In the immediate neighborhood of the cathode, sodium ions are attracted by the negative charge, and each ion acquires a single electron, becoming an atom of sodium metal. See Figure 15-5. In terms of the restriction just mentioned, a chloride ion must leave the region of the cathode at the same instant, or else another sodium ion must move in from the liquid further out. At the anode a similar process occurs: a chloride ion, pulled in by the positive charge, loses an electron and becomes a chlorine atom (to combine an instant later with a brother atom into a molecule), while a sodium ion moves away or another chloride ion moves in.

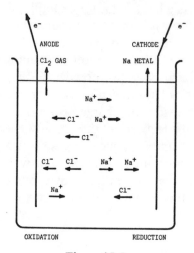

**Figure 15-5**

In the bulk of the liquid, then, two motions are going on at the same time: a stream of positive ions is passing from the anode to the cathode, at the same time as a similar stream of negative ions goes from the cathode to the anode. At every point, either around the electrodes themselves or anywhere else in the liquid, exact electrical neutrality is preserved.

A second restriction in any electrolytic reaction is the obvious but important one that for every electron that the cathode loses, one electron must be supplied from the anode. The total number of electrons in the external circuit cannot change.

At the cathode, sodium ions are being reduced, while chloride ions are being oxidized at the anode. Chemists have now very nearly agreed to standardize the meanings of the words anode and cathode. The *anode* is always the pole at which *oxidation* occurs, and the *cathode* is the one where *reduction* occurs*. The cathode is therefore the electrode into which electrons flow from the external circuit, and the anode the one *from* which they flow to the external circuit. This will be the case even when the cathode is labeled "positive" and the anode "negative."

This very simple type of electrolysis is of considerable commercial importance. The sodium and chlorine from salt are both valuable. Two very important structural metals, aluminum and magnesium, are made by electrolysis of their molten compounds. Aluminum is made by electrolyzing aluminum oxide, bauxite, dissolved in another molten mineral, cryolite. Magnesium chloride from sea water is the source of magnesium.

In every electrolysis we are dealing with two separate reactions, one of them occurring at each electrode. Yet these two reactions are, in another sense, inseparable: neither can happen unless the other happens at the same time. Consequently, they are usually spoken of as *half-reactions*. In the case of sodium chloride, the half-reactions are:

$$\text{Anode reaction: } 2Cl^- \longrightarrow Cl_2(g) + 2e^-$$
$$\text{Cathode reaction: } Na^+ + e^- \longrightarrow Na.$$

Since we'll return to it again and again, it is worthwhile to summarize the basic material we have covered so far.

(a) Ions move. In general, an ion will be attracted to the electrode having the opposite charge.

(b) Electrical neutrality is maintained within every tiny volume of the liquid: as positive ions leave, either negative ions must also leave, or other positive ions must enter.

(c) Redox reactions always occur: oxidation at the anode, reduction at the cathode. These can be presented as half-reactions.

(d) The number of electrons used in the reduction at the cathode must be exactly equal to the number set free by the oxidation at the anode.

*Electrolysis: Aqueous Solution of Sodium Sulfate*

We now move to a much more complex case. This one is easy to demonstrate. A solution of sodium sulfate is prepared, roughly

---

*Unfortunately, some textbooks still fail to follow this practice, and serious confusion often results, especially when dealing with voltaic cells, in which the anode may be labeled "negative."

half-molar, containing a pair of convenient mixed indicators. I use thymol blue and methyl red, both of which are yellow in the neutral range. The thymol blue turns blue above pH 8 or so, and the methyl red is red below about pH 6. The mixture is put into a large U-tube and platinum electrodes are introduced. See Figure 15-6. Bubbles of gas immediately start to form: hydrogen at the cathode and oxygen at the anode, as most eighth-graders know. But at the same time, the solution around the anode turns red, and that around the cathode, blue. As the demonstration continues, the red and blue regions extend down the arms of the tube, until they meet at the bottom.

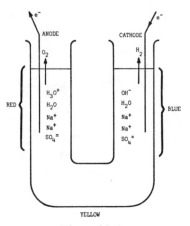

**Figure 15-6**

At any time, the electrodes can be removed, and the solution poured into an empty flask, where the yellow color is restored. If the blue side is poured first, the solution in the flask remains blue until the last drop from the U-tube goes in (sometimes the tube must be rinsed with the solution). This shows clearly that exactly equal numbers of hydronium and hydroxide ions are being produced.

This reaction, producing hydrogen, oxygen, acid, and base, occurs when the electrolyte has as its cation (positive ion) hydronium, or one of the more active metals, such as sodium, zinc, or iron, and the anion is that of an oxy-acid: sulfate, nitrate, phosphate. In these cases, *neither* ion of the electrolyte is altered in the process. Water alone undergoes a change.

At the anode, the red color of the indicator shows that

hydronium ions are produced in the liquid in addition to the oxygen gas that is evolved. At the cathode, as hydrogen is freed, the solution is made basic by hydroxide ions. The standard way to treat these reactions is to give the following half-reactions:

Anode: $2H_2O \longrightarrow O_2(g) + 4H^+ + 4e^-$ (omitting, as I shall throughout the rest of this chapter, the water molecules to which the proton is attached in the hydronium ion)

Cathode: $2H_2O + 2e^- \longrightarrow H_2(g) + 2OH^-$.

This is a straightforward description of what occurs. Notice that in order to keep the balance-sheet of electrons straight, twice as many water molecules must be reduced at the cathode as are oxidized at the anode. This explains the famous two-to-one ratio of hydrogen to oxygen in the entire process.

The only objection to this conventional description of the half-reactions is that it makes no particular sense. *Why* does water behave so differently at the two electrodes? What happens to the sodium ions and the sulfate? And what has the whole business to do with other chemical principles? I prefer, therefore, to handle the processes occurring at the electrodes in the following way.

In the solution, there are actually five species: large numbers of sodium and sulfate ions, water molecules, and tiny numbers of hydronium and hydroxide ions. Since the cathode presents the simpler problem, we'll deal with it first.

Both sodium and hydronium ions will be attracted toward the cathode. Of these the sodium ion requires a higher voltage to be reduced than does the hydronium (more electron pressure; the reason for this will be seen later in this chapter). Therefore it is the hydronium that accepts electrons:

$$2H^+ + 2e^- \longrightarrow H_2(g).$$

The loss of these hydronium ions upsets the equilibrium of the ionization of water:

$$H_2O \rightleftharpoons H^+ + OH^-.$$

By Le Chatelier's Principle, as the hydronium is removed, more water immediately ionizes. Thus, for each hydronium ion reduced, one hydroxide ion is set free in the solution. See Figure 15-7. Since these hydroxide ions have been virtually "created" in the neighborhood of the cathode, we must take care of the requirement of electrical neutrality in the liquid. This could be achieved in three ways: (a) one sodium ion could move into the neighborhood as each hydroxide is produced; (b) sulfate ions could move away, one sulfate for each two hydroxides; (c) the hydroxide ions could themselves move away. In fact, all three of these things happen. The solution therefore becomes essentially a sodium hydroxide solution (and this process is indeed used in the manufacture of the lower grades of sodium hydroxide, except that sodium chloride is used instead of the sulfate).

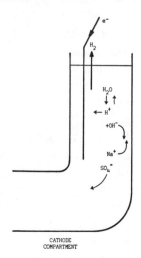

**Figure 15-7**

At the anode, things start out in similar fashion, but my explanation will be much more hypothetical. Suppose a hydroxide ion is oxidized:

$$OH^- \longrightarrow (OH + e^-).$$

I put this reaction in parentheses, because the neutral OH group would be what is called a "free radical": a grouping with tremendous reactivity. Pairs of hydroxyl radicals would instantly combine, with the overwhelming probability that they would line up in the position OHOH; i.e., with the positive end of one attracted to the negative (oxygen) end of the other. The enclosed hydrogen would then shift to the right-hand OH group—this is of course a Brønsted acid-base reaction:

$$OH + OH \longrightarrow O + H_2O.$$

The resulting oxygen atoms would then combine in pairs to form molecules of oxygen gas.

In the surrounding solution, the same type of ionic activity occurs as at the cathode: as the hydroxide ions are removed, the ionization equilibrium of water is shifted to produce excess hydronium ions, sodium ions move away, and sulfate ions move in. See Figure 15-8. The net result of this would be eventually a solution of sulfuric acid around the anode. I scarcely need to note that this would not be a practical way to make sulfuric acid.

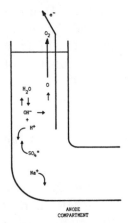

**Figure 15-8**

I should emphasize the *weaknesses* in the mechanisms suggested here for the reactions at the two electrodes. It has been pointed out that the concentrations of hydronium ion (at the cathode) and of hydroxide (at the anode) are so small that reactions involving these two ions are relatively unlikely. Yet the ionization of water occurs at such a rapid rate that even though concentrations of these ions are small, the "supply" is rapid and constant. Also, exactly these reactions *must* occur (a) at the cathode during electrolysis of a strong acid such as sulfuric, and (b) at the anode when sodium hydroxide is electrolyzed. I prefer, then, to use this mechanism for my own clear understanding of what occurs, and to make it comprehensible to students.

One final matter needs to be considered before we leave the sodium sulfate solution. If ultimately all that occurs is the electrolysis of water itself, why must we use an electrolyte in solution? Why bother with the sodium sulfate? The answer to that question ties in with the requirement of electrical neutrality in the liquid. Suppose we did try to electrolyze pure water, with a 6-volt current, let's say.

It would work. A very tiny trickle of current would pass: something like $10^{-7}$ amperes with electrodes of area 1 cm² placed 10 cm apart, yielding a few billionths of a mole of products per hour. As each hydronium or hydroxide ion is produced, by oxidation or reduction of its companion ion, in the water around the electrodes, ions of opposite electrical charge must be available to maintain electrical neutrality. Hydronium or hydroxide ions cannot simply "pile up" around the electrode; and the ionization of water itself provides quantities of these ions far too small to make the process efficient. The function of the sodium sulfate (or other electrolyte) is to provide these extra ions which can move in or out, as needed, to balance the new ions that are created.

## *Electrolysis of Aqueous Solutions with Inert Electrodes: General*

The two cases of electrolysis covered thus far illustrate the simplest (molten salt) and the most complex (sodium sulfate solution) cases of electrolysis with inert electrodes. There are a number of "mixed" cases that merit a little attention.

In discussing the sodium sulfate solution, I said that the sodium ion could be neglected because more voltage was needed to reduce it than to reduce the hydronium ion. However, if copper or silver sulfate solutions are electrolyzed, it will be the metallic ion, rather than the hydronium, that is reduced. Copper or silver will "plate out" on the cathode. An "activity series" will be found in most texts, usually as a table of oxidation potentials.

Voltage figures are often given in these tables, whose meaning we'll consider shortly. But any metal with a negative oxidation potential will deposit in preference to hydrogen gas: it takes less energy to reduce these metals than to reduce the hydronium ion.

| *A Condensed Activity Series* |
| :---: |
| Metals with positive oxidation potentials |
| sodium |
| magnesium |
| zinc |
| iron |
| tin |
| Metals with negative oxidation potentials |
| copper |
| mercury |
| silver |

Thus, in a solution of the sulfate or nitrate of these less "active" metals, the *cathode* reaction will be the same as if water were not present—as though the molten salt were being used. The *anode* reaction will be the same as with sodium sulfate or nitrate; that is, oxygen will be produced, along with hydronium ion in solution. See Figure 15-9. If current is passed for enough time through a copper sulfate solution, all the copper ions will be removed, to be replaced by hydronium ions, and the final solution will be the same as if sulfuric acid had been added to water. Towards the end of such an experiment, as the copper concentration drops to very low values, both hydrogen and copper metal will be set free at the cathode. Throughout the process, there will be a steady drift of sulfate ions toward the anode.

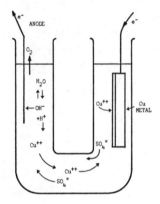

**Figure 15-9**

On the other hand if an aqueous solution of sodium chloride is electrolyzed, the cathode reaction will be the same as with sodium sulfate solution; that is, hydrogen gas and hydroxide ions will appear there. But at the anode, chlorine gas will be formed. The "rules" for the anode reaction are less clear-cut than those for the cathode. For simplicity it can simply be assumed that if a halide ion (chloride, bromide, or iodide) is present in water solution, the halogen will be formed instead of oxygen. Other cases (such as nitrites, sulfites, etc.) need not be considered here. A potassium iodide solution provides a good demonstration, with the brown color of the iodine very obvious in the anode compartment. See Figure 15-10.

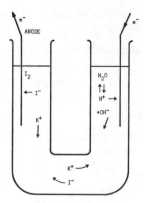

**Figure 15-10**

## *Aqueous Solutions with an "Active" Anode: Metal-Plating*

The last case of electrolysis to be considered can be demonstrated by using copper instead of platinum electrodes in a sodium sulfate solution. In this case, hydrogen and hydroxide ion appear at the cathode, but there is no gas at the anode. If the process is carried on long enough, a pale blue color will appear around the anode. This can be accentuated by adding ammonia water, whereupon the deep blue of the cuprammonium complex shows up.

This reaction will happen with any anode that is more "active" than the noble metals, gold and platinum. The positive metal ions themselves break loose from the metallic matrix, and it is thus the metal of the electrode that is oxidized at the anode rather than any ion in the solution.

The set-up I have described is fine for a demonstration using copper as the anode in a sodium sulfate solution—the metal of the cathode is not important. But in commercial practice, the active anode is used in *metal-plating,* and the electrolyte used is a salt of the metal. Suppose an iron spoon is made the cathode, a chunk of copper the anode, and they are put into a solution of copper sulfate. See Figure 15-11. At the *anode,* copper will be oxidized and goes into solution as $Cu^{++}$ ions. At the *cathode,* copper ions will be reduced and plate the spoon with copper metal. And in the solution, sulfate ions will stand essentially still, serving simply as a negatively charged "railroad track," maintaining neutrality of

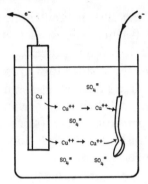

**Figure 15-11**

charge as the positive copper ions move from anode to cathode. The same process is used for plating with all kinds of inactive metals, including silver for tableware.

One of the most interesting applications of electroplating, though, is in the large-scale refining of copper. The principal use of metallic copper is for electrical wire, and tiny amounts of impurities greatly increase the resistance of the wire. But as copper comes from the smelters (usually at the site of the copper mine), it is usually only 97-99% pure, containing iron, zinc, silver, arsenic, and sometimes gold and platinum.

Slabs of this impure copper (called "blister copper") are made the anodes, with thin copper sheets as the cathodes, and copper sulfate as the electrolyte. See Figure 15-12. The voltage is carefully regulated: kept just high enough to oxidize the copper at the anode. The copper is, of course, oxidized, but so also are the more active elements, such as iron, lead, zinc, and arsenic. The precious metals, silver, gold, and platinum, are not oxidized and

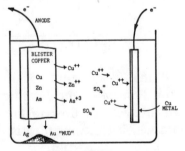

**Figure 15-12**

drop to the bottom of the tank as the "anode mud." This precious mud is collected and further refined, sometimes having enough value to pay for the whole refining process.

At the cathode, only the copper ions are reduced, since this is the ion of the least active metal in solution. The other metallic ions remain in solution, and their presence gradually decreases the concentration of copper ions there (since the total charge of positive ions remains constant). This solution, containing the more active metals, is eventually discarded.

In this process, as with more conventional electroplating, the sulfate ions serve as a kind of "track" across which the copper ions can move from anode to cathode.

### The Voltaic or Galvanic Cell

At the beginning of this chapter I mentioned Volta's "battery," made of zinc and copper with paper moistened with electrolyte solution. With a good galvanometer, Galvani's reaction can be easily demonstrated. Take a beaker of salt solution (any salt, but sodium or potassium nitrate are convenient ones), connect any two metals to the galvanometer, and dip them both into the salt solution. See Figure 15-13. Metals like a nickel and a dime or a nickel and a penny work well, or use a steel screwdriver and a piece of copper wire.

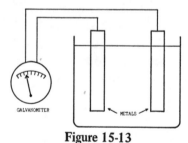

**Figure 15-13**

It will be found that the current sometimes flows one way, sometimes the other. Remember the definitions adopted here: the electrode *from* which electrons flow to the galvanometer is the anode; the one into which electrons flow is the cathode. A given metal strip may act as either anode or cathode, depending on the other metal used with it.

After establishing which direction the needle moves for a given direction of electron flow, a very simple "activity" series can be set up, where the more "active" metal of a pair is the one that acts as the anode. That is, zinc is more active than copper, and electrons flow in the wire from the zinc strip to the copper strip. But copper is more active than silver. A good group of metals, usually available in the lab, is silver, copper, zinc, iron, lead, and magnesium.

In explaining the action of a voltaic cell, the conventional treatment starts with the fact that current is produced by oxidation and reduction reactions at the electrodes, and then goes on to study the details of the reactions. As a scientist, I must insist on something more fundamental, even though some speculation is needed: "proofs" are hard to come by in what I'm about to say.

Let us make this assumption, then. If a piece of metal—any metal—is dipped into water (preferably containing an electrolyte), the first thing that happens is that some metallic ions break loose from their matrix (as with the "active anode" in metal-plating), escaping *as ions* into the solution. See Figure 15-14. The cause of this initial escape, fundamentally, is the drive toward increased entropy: the atoms are jiggling in random fashion, and some of those on the surface break free of the attraction of the "electron cloud" that binds the ions together. The escape of these ions is comparable to the dissolving of a crystal of sodium chloride, except that only positive ions escape. These ions are then free to wander—temporarily—through the liquid.

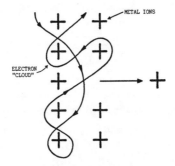

**Figure 15-14**

But as each positive ion breaks free, a net negative charge is left on the metal; that is, one excess electron remains behind, to circulate throughout the strip. See Figure 15-15. As more and more ions escape, this charge rises. But a negatively charged piece of metal will attract positive ions from the solution: it will act as a cathode. So very shortly (in a tiny fraction of a second, in fact), positive ions will begin to return from the liquid to the metal. Presently, an equilibrium will be established. Metal ions are escaping, with entropy gain, as fast as they return, with energy loss as opposite charges come together. The overall process will be closely analogous to the vapor pressure situation that we've referred to again and again throughout this book. And here, as there, the equilibrium can be shifted in predictable ways, using Le Chatelier's Principle or collision theory.

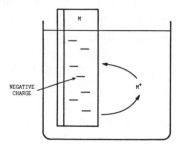

**Figure 15-15**

*The Daniell Cell*

One of the best ways to demonstrate and explain the action of a standard voltaic cell is to set up the equipment shown in Figure 15-16, a clumsy but transparent form of the Daniell cell. The bottom of a U-tube is filled with sand, used simply to prevent ready mixing of the liquids in the two arms. Then, with a cork in the right arm of the tube, the left arm is filled with zinc sulfate solution. Reversing the cork, fill the left arm with copper sulfate solution. (It is important to fill this cell in the above order; zinc ion on the copper side doesn't matter, copper ion on the zinc side is harmful.) Now remove the corks, and put a strip of the corresponding metal into each arm of the tube. The resulting cell won't light a flashlight bulb, but will show strong effects with a voltmeter or ammeter.

When the metals are connected by a wire, there will be a steady

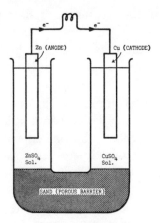

**Figure 15-16**

flow of current—electrons flowing from the zinc to the copper electrodes. That is, the zinc is the anode, copper the cathode.

Consider now what happens in each arm of the tube *before* the electrodes are connected. In each case, the equilibrium I described above is quickly established. Each strip gains a negative charge (this is why I have definitely *not* followed the confusing convention of labeling the electrodes "positive" and "negative") as ions escape, and its ions, already present, return from the surrounding solution. Thus we have two negatively charged poles, but the zinc has a higher negative charge than the copper. The factors that control the size of the negative charge will be considered shortly (see p. 223).

Now connect the electrodes. This immediately disturbs the equilibrium at each. The higher charge of the zinc means that its electron cloud has higher pressure than that on the copper, so electrons flow from zinc to copper. The charge on the zinc strip, now reduced, cannot pull positive ions from the solution as quickly as it did, but the rate of escape of zinc ions remains unchanged. Consequently, the metal itself begins to disintegrate. At the copper pole, the higher-than-normal charge of the metal pulls copper ions from solution more quickly than at equilibrium, and this metal, like any cathode, begins to be "copper-plated." In fact, at each pole we have exactly the situation that we've already explored in the case of electrolysis with an "active anode." The only difference is that the energy for the reaction is provided by the tendency of the zinc to ionize, rather than coming from an external source.

From here on, the analysis of the cell is fairly simple, requiring no new principles. We must satisfy our basic restrictions: (1) there must be oxidation at the anode and reduction at the cathode; (2) in the solution, the total number of positive and negative charges must be equal; and (3) every electron that leaves the metallic part of the circuit must be replaced by one entering it.

For the first point, zinc is oxidized as its ions escape, and copper ions are reduced. See Figure 15-17. To maintain neutrality in the solution, as each zinc ion enters, *either* a sulfate ion must move toward, or a zinc ion away from, the zinc strip. And in the copper compartment, as each copper ion is lost, it must be replaced by another positive ion, or a sulfate ion must move away. In the bulk of the solution, then, there will be steady motion of both zinc and copper ions toward the right (in this diagram), and of sulfate ions toward the left. The total quantity of sulfate ions in the solution will not change, but with time the total amount of copper will drop, and that of zinc will rise.

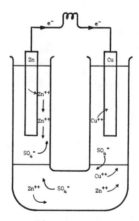

**Figure 15-17**

To satisfy the final restriction, electron balance, the loss of zinc ions from the anode will leave excess electrons which move through the external circuit to the cathode. There they are used to reduce copper at the cathode, an atom of copper for each atom of zinc.

One question crops up: What does the sand, the porous barrier, do? Within each arm of the U-tube, there is, of course, relatively free movement of all ions and molecules. But in the sand layer, with its narrow channels, this movement is restricted. In this layer there can be steady flow of zinc ions together with any copper ions to the right, along with flow of sulfate ions to the left.

This is an important consideration. The zinc ions that flow into the cathode compartment have no effect on the reaction there, since the ion that is reduced at the cathode is the one with lowest activity (in the refining of copper, the more active metals, including zinc, were not reduced at the cathode). So the zinc ions mix harmlessly with the copper ions in the cathode compartment.

But if copper ions were to stray into the anode compartment, the cell would become less effective. Its action depends on the fact that metallic ions leave the strip faster than they return. But if copper ions were present, they would immediately "plate out" on the zinc (as anyone knows who has dipped a strip of zinc into copper sulfate solution). This would tend to "short circuit" the cell—details of this action will be considered when we take up corrosion.

There will be no change in the reactions of a Daniell cell as long as the zinc metal holds out at the anode, and the copper ions last in the cathode compartment. Daniell cells were very important in the era before electric generators were in large-scale use, and there were a number of ingenious modifications of the basic principle. In the first place, my clumsy U-tube with a sand barrier was never used. The porous barrier was sometimes a cup of unglazed pottery. A thin sheet of copper was placed in concentrated copper sulfate solution outside of this, and a heavy chunk of zinc, in weak zinc sulfate solution, was put in the porous cup. See Figure 15-18.

Such "wet cells" were reasonably effective. But in telegraph offices as recently as the second decade of this century, one saw banks of so-called "gravity cells." In a gravity cell, the porous cup was done away with. The two solutions were kept separate simply by the difference in their specific gravities. See Figure 15-19. In the lower part of the jar was a *saturated* solution of copper sulfate, its strength maintained by large chunks of copper sulfate crystals. On top of this solution, was floated a weak solution of zinc sulfate. The electrodes were so-called crow's feet of zinc and copper. In this cell, upward diffusion of copper ions was kept

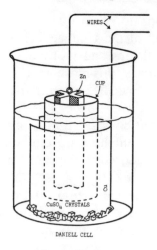

**Figure 15-18**

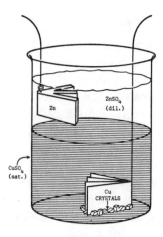

**Figure 15-19**

small by the higher density of the copper sulfate solution. But this in itself would be inadequate. The separation was maintained primarily by the one-way migration of positive ions. As long as the cell was being used, positive ions could move only from the anode toward the cathode; that is, downward. In an old-fashioned telegraph line, the circuit must remain closed; that is, the battery must be operating at all times. Consequently, there was always a sharp line dividing the colorless upper solution from the blue lower one.

Copper ions plating out of the lower solution were replaced by the dissolving of copper sulfate crystals, whose sulfate ions migrated toward the upper layer. Eventually, the density of this upper layer would increase to the point where the solutions would physically mix. A battery must be refilled before this happens.

## Oxidation Potentials, and the Voltage of a Cell

In this chapter, I have referred several times to the "activity" or oxidation potential of metals. It is time now to look into this more carefully.

I said that if a metal strip is put into water (or an electrolyte solution), it acquires a negative charge, this charge being ultimately the result of entropy gain. The size of the charge in a given case depends on three factors. The first is the energy needed to pull an ion free of the surrounding electron cloud. This is roughly proportional to the *ionization energy*; i.e., the energy needed to remove electrons from a free metal atom. The second factor is *hydration energy* of the metallic ion: the energy released when a free metallic ion attracts, and is surrounded by, the highly polar molecules of water. The third factor involves the equilibrium between escaping and returning ions: the higher the concentration of a metal ion already present in the surrounding solution, the lower the charge needed to pull ions back as fast as they escape.

In order to measure an oxidation potential, two electrodes must be in contact with the solution. This means that two equilibria are always involved. For this reason, oxidation potentials are never absolute figures—an arbitrary zero point had to be established. This zero point is now defined as the potential of a "hydrogen electrode" under specified conditions. See Figure 15-20. The electrode is a piece of spongy platinum metal saturated with hydrogen gas at 1 atmosphere pressure and 25°C. The hydrogen takes the place of the metal of an electrode and is in equilibrium with hydronium ions ("hydrogen ions") at 1M concentration.

To establish the oxidation potential of a particular metal, let's say zinc, the equipment shown in Figure 15-21 can be used. The cell is the same in principle as the Daniell cell, and the solutions are 1M hydrochloric acid and 1M zinc chloride. The cell is connected to some type of nul-current voltmeter (potentiometer, etc.). In such a cell, the voltage measured is the *standard oxidation*

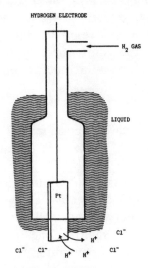

**Figure 15-20**

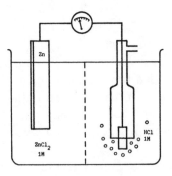

**Figure 15-21**

*potential* for the reaction, and in this case will be 0.76 volts, with the zinc as anode.

The half-cell reaction for the zinc is:

$$Zn \rightleftharpoons Zn^{++} + 2e^-,$$ ox. potential +0.76 v. The voltage figure is given a positive sign (again arbitrarily) which means that in this cell the zinc is oxidized. The other reaction in the cell is:

$$2H^+ + 2e^- \rightleftharpoons H_2(g),$$ whose potential is by definition 0.00 volts. The hydronium ions are reduced in this reaction.

If in the same type of cell the metal were copper, with 1M

copper chloride the electrolyte, the copper will be the cathode. The reaction is written:

$$Cu \rightleftharpoons Cu^{++} + 2e^-,$$ with a voltage of $-0.34$, which means that in this cell the reaction proceeds from right to left as copper ions are reduced to copper metal.

Notice that in the zinc-hydrogen cell the hydrogen electrode is the cathode, while in the copper-hydrogen cell it is the anode.

Since the voltage figures represent ionization tendencies on an accurate relative scale, they can be combined. In a zinc-copper cell (a Daniell cell) with 1M solutions, the total voltage will be the algebraic sum of the half-reaction voltages when each is written in the direction in which it actually goes:

Zn $\longrightarrow$ Zn$^{++}$ + 2e$^-$ +0.76v

Cu$^{++}$ + 2e$^-$ $\longrightarrow$ Cu +0.34v

Overall voltage 1.10.

Similar figures can be used for any combination of metals. A magnesium-silver cell, for example, would provide 3.17 volts in the standard cell, with oxidation potentials for the magnesium of +2.37 and for silver of $-0.80$.

The concentration of ions in solution obviously affects the equilibrium around a metal electrode, and therefore the voltage of a cell. The explanation of the quantitative effects of concentration requires a grasp of thermodynamics beyond our scope here. But the qualitative effects are clear enough: as the concentration of metal ions in the anode compartment decreases the voltage rises, since ions can escape more rapidly than they return. In the cathode compartment, lower concentration of ions means *lower* voltage, since fewer ions already present are available to return*.

*The quantitative formula for the calculation of concentration effects is the *Nernst Equation:*

$$E_{298} = E^o_{298} - 2.303\frac{RT}{nF}\log\frac{[Zn^{++}]}{[Cu^{++}]}.$$

Here $E_{298}$ is the ideal voltage of an actual cell at 298°K, $E^o_{298}$ is the standard oxidation potential, R and T have their usual meanings, $F$ is the number of coulombs in a faraday (a mole of electrons), and n is the number of electrons transferred per ion oxidized or reduced.

For the Daniell cell at 25°C, where n = 2, this equation becomes:

$$E_{298} = 1.10 - 0.0296(\log\frac{[Zn^{++}]}{[Cu^{++}]}).$$

This equation gives quantitative expression to the idea of this paragraph. As [Zn$^{++}$] rises or [Cu$^{++}$] falls, the value of the negative term increases, and the voltage goes down. When the concentrations of the metal ions are equal, the negative term vanishes and the voltage is the value given by the standard oxidation potentials.

*Corrosion of Metals*

Most of us first made hydrogen gas early in an elementary chemistry course by adding a strong acid, such as hydrochloric, to zinc metal. But a batch of zinc sometimes turns up that gives off hydrogen only very slowly. Lab manuals then suggest that a little copper sulfate solution be added to the acid. When this is done, the reaction picks up beautifully. The instructor's only problem is then to explain to his neophyte students *why* this works. My answer, at the start of the year, is simply that the copper solution "acts as a catalyst"—not strictly accurate, but one of those half-truths that must be used in the early stages of any course.

The explanation is actually much more involved. If a piece of pure zinc is dipped in strong acid, the first reaction is the one we've been dealing with in connection with galvanic cells. Zinc ions escape, a negative charge builds up on the metal, and equilibrium is established, with the zinc strip closely surrounded by a thin layer of zinc ions, escaping from, and returning to, the metal. See Figure 15-15. Nothing further can happen in this case. But if a second, less active metal, such as copper, is in the solution and in contact with the zinc, the zinc dissolves rapidly and hydrogen gas appears *on the copper.*

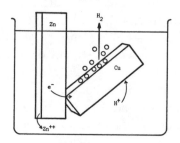

**Figure 15-22**

If very pure zinc is available, this can be demonstrated easily. Dip the zinc strip in the acid and nothing happens. Put a piece of copper in the acid (again with no reaction) and press the zinc against it, either above or below water level. Bubbles immediately form on the copper and break free (Figure 15-22). Since pure zinc strips are hard to find except by accident, the effect can be simulated by first "amalgamating" the zinc with mercury. A small amount of mercury is rubbed on the surface of the strip, making it shiny. In the amalgam the mercury acts only as a carrier for the zinc, whose action is then similar to that of the pure metal.

It is obvious that we have here a voltaic cell. The electrolyte is hydrochloric acid, and the reactions at the electrodes are:

At the anode: $Zn \longrightarrow Zn^{++} + 2e^-$

At the cathode: $2H^+ + 2e^- \longrightarrow H_2 (g)$.

The overall reaction is the one commonly written for the generation of hydrogen:

$Zn + 2H^+ \longrightarrow Zn^{++} + H_2$, and the copper is indeed only a catalyst, remaining unchanged.

But when copper sulfate is used in the laboratory process, it *does* undergo a change—an electrolytic reaction. In the first instance, as zinc ions escape, the less active copper ions can return from the solution, forming a somewhat porous coat of copper metal. Now, at each point where the copper touches zinc, a tiny cell is set up, and hydrogen is formed, as before, on the copper. It is this reaction, incidentally, that makes it important to allow no copper in the anode compartment of a Daniell cell.*

The complications in this simple experiment, one of the earliest in a chemistry course, have a direct bearing on the complex subject of corrosion in general. Early in the history of metallurgy it was known that "wrought iron" was slow to rust, whereas "cast iron" would quickly rust away when it was exposed to moisture. Anchor chains were made of wrought iron until recent times.

Cast iron is the crude metal as it comes from a blast furnace, where carbon had dissolved in the molten iron. As the liquid metal cools, carbon crystallizes from solution, leaving a metallic surface like that of the copper-treated zinc, with bits of carbon in contact with the iron metal. When the metal is wet, voltaic cells are set up all over its surface. Each carbon particle can act as a cathode, and iron ions escape to form rust.

Nowadays, the metallurgy of steel is so advanced that the old term "wrought iron" is seldom used. This was iron that had been

*The question arises, of course: *Why* can *copper* ions penetrate the screen of zinc ions that surrounds the metal, while hydronium ions of the acid do not? The answer, I think, is that hydronium ions *do* penetrate, initially, to form a very thin layer of gas. This, a non-conductor, prevents further motion of ions or electrons. This phenomenon is called "polarization" of the electrode. In the case of the most active metals, such as sodium, polarization is inadequate to prevent further reaction, and the metal will ionize rapidly, setting free hydrogen gas. I have wondered, though, if pure enough sodium could be prepared so it would *not* react with water!

worked, at high temperatures in a semi-plastic state, until most of the carbon was oxidized. The principal impurities in this metal were fibers of non-conducting slag, and the metal behaved like the pure zinc, corroding only very slowly.

The galvanic action of corrosion is well illustrated by two ordinary household items, "tin" cans and galvanized iron pails. See Figure 15-23. Tin cans are made of steel with a thin outer coating of tin. Tin is a relatively inactive metal, and as long as the tin-coating is intact it suffers very little corrosion. As soon as the coating is injured, however (by denting, or after the can is opened), moisture will produce a galvanic cell, in which the tin is the cathode, and the iron, as anode, ionizes. So tin cans on a trash-heap will quickly rust away.

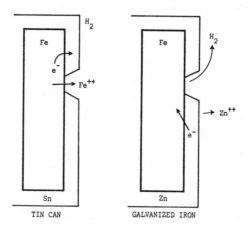

**Figure 15-23**

Galvanized iron, on the other hand, has zinc, the more active metal, as the coating on the iron. When this coating is broken and a cell is formed, the zinc is the element that ionizes, and the main structural element, iron, is protected from corrosion. The zinc in this case is called the "sacrificial" metal.

Sacrificial protection of this sort is routinely used on metal boats. Chunks of active metal, such as magnesium, are bolted to the metal hull of the boat. During the corrosion process, a steady flow of electrons to the steel hull of the boat prevents breakdown of the structural metal, and the wasted magnesium pieces are easily replaced.

## The Lead-Acid Storage Battery: Discharging

One of the most commonly used cells is the ordinary automobile storage battery. The chemistry of this is ingenious, and not
as complicated as it sometimes sounds. In a "charged" battery the
situation is the one shown in Figure 15-24, with the anode
essentially metallic lead, and the cathode a thick coating of lead
dioxide, $PbO_2$, on a base of lead metal. The electrolyte is dilute
sulfuric acid. The basis for operation of the cell is the oxidation of
lead at the anode (like the zinc in a Daniell cell), and the reduction
of the $Pb^{+4}$ ion to $Pb^{+2}$ at the cathode. What makes the cell differ
so greatly from simple galvanic cells is that lead sulfate, $PbSO_4$, is
very insoluble, and therefore the $Pb^{++}$ ions around each electrode
remain in place instead of moving through the liquid.

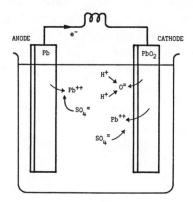

**Figure 15-24**

When the cell is discharging—doing work—lead ions are produced from the metal at the anode, and electrons are pumped
through the external circuit to the cathode, where they reduce the
lead oxide:

Anode reaction: $Pb \longrightarrow Pb^{++} + 2e^-$

Cathode reaction (simplified): $Pb^{+4} + 2e^- \longrightarrow Pb^{++}$.

At the anode, the lead ions immediately combine with sulfate
ions that come from the solution, and lead sulfate is deposited on
the electrode. At the cathode, the divalent lead ions *also* combine
with sulfate ions, and there too lead sulfate is deposited. But,
simultaneously, the oxide ions of the lead dioxide are set free
from their combination with lead. The oxides hook up with

hydronium ions of the acid to form water (I am oversimplifying, of course), and the economics of this—

$$20^= + 4H^+ \longrightarrow 2H_2O$$

—means that for each lead ion oxidized at the anode, and reduced at the cathode, two sulfate ions and four hydronium ions are removed from the solution, to be replaced by water. Since these are just the proportions that exist in dissolved sulfuric acid, the effect is to make this solution more dilute as the battery discharges. The state of charge of the battery can then be determined by measuring the sulfuric acid concentration with a hydrometer.

## Charging the Battery

Assume a completely discharged battery (a state not desirable, for practical reasons). Each electrode would then be covered with lead sulfate on the lead metal base. To charge the battery, current is supplied from a generator, in the opposite direction from the current that flowed during discharge. See Figure 15-25. The

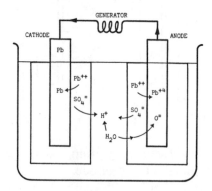

**Figure 15-25**

original anode is made the cathode. There must still be some sulfuric acid in the solution to facilitate movement of ions. At the new cathode, lead sulfate is reduced to lead metal, freeing sulfate ions to return to the solution. As electrons are pumped out of the new anode, the divalent lead of the lead sulfate is oxidized to $Pb^{+4}$, which combines with oxide ions of water (again an over-simplification), setting free both sulfate and hydronium ions:

$$\text{Cathode: } Pb^{++} + 2e^- \longrightarrow Pb$$
$$\text{Anode: } Pb^{++} \longrightarrow Pb^{+4} + 2e^-$$
$$\text{and } Pb^{+4} + 2H_2O \longrightarrow PbO_2 + 4H^+.$$

Considering the overall economy of the cell, the result is to restore each electrode to its "charged" state and add sulfuric acid to the electrolyte solution.

## Conclusion

This treatment of electrochemistry has barely scratched the surface of an important field. The new storage batteries, and the many forms of "dry cell" now available have revolutionized the practical use of battery-operated equipment. In "fuel cells," very energetic combustion reactions, such as the combination of hydrogen and oxygen, can be harnessed to produce electron flow directly. Fuel cells thus cut out the inefficiencies of standard power-plant processes, where the chemical process produces heat energy, which is converted to mechanical energy, which finally moves electrons in a generator. Fuel cells may well provide the power for the non-polluting automobiles of the future.

# 16

## Dynamic Equilibrium in Small, Oft-Repeated Steps: Serial Processes

An old story from farm country has a storekeeper selling eggs at 34¢ a dozen. Somebody says, "What do you pay for 'em?" "Thirty-five cents." "You can't make money that way!" "Oh, but I sell so many of 'em!"

That isn't quite the way the processes work that we're going to deal with here. But reverse the figures, and you have it. If a process can show a very slight profit, chemically speaking, then it is usually possible to devise a way to do it often enough to make it highly efficient.

### *Distillation*

In earlier chapters, when we dealt with vapor pressure and its relation to boiling, I was careful to use cases where the solute was non-volatile, so we had only the vapor pressure of the *solvent* (usually water) to consider. The situation becomes a good deal more complicated when the solute is volatile too. Here I'll consider only the simplest case of this sort, one where the two components of the mixture are completely soluble in one another. Specifically, I'll use methanol (methyl alcohol) and water.

232

We have seen that the boiling point of a liquid is the temperature at which its vapor pressure equals the pressure of its
surroundings. If something is boiled in a container exposed to the
air, the vapor pressure must equal atmospheric pressure. Then as
the liquid boils, escaping molecules push back the air, and move
completely away from the surface of the liquid. If the escaping
vapor is condensed, we have a distillation process. See Figure 16-1.

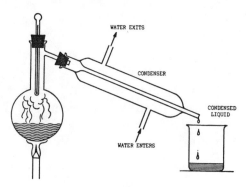

**Figure 16-1**

In Chapter 6 we saw that when a solute is present in water it
decreases the vapor pressure of the water at any given temperature. The final vapor pressure is proportional to the mole fraction
of water in the mixture. This is Raoult's Law, $P_W = P_{wo}M_W$,
where $P_{wo}$ is the vapor pressure of *pure* water at the temperature
in question, and $M_W$ is the mole fraction of water present. If 1
mole of glycol (non-volatile) is added to 4 moles of water, the
mole fraction of water is 4/5 or 0.8; and the vapor pressure of the
solution at any temperature is four-fifths of that of pure water.

The reasoning applies even if the solute is volatile. But in that
case, the total vapor pressure will be made up of this reduced
water vapor pressure *plus* whatever pressure the vapor of the
solute adds. So if instead of glycol we add 1 mole of methanol to
4 moles of water, the vapor pressure due to the water is still
four-fifths of the pressure for pure water. But if the vapor
pressure of the methanol is added, we can use Raoult's Law to
figure this. It would be one-fifth of the vapor pressure of pure
methanol at the temperature used.

At standard pressure, methanol boils at about 65°C, and water at 100°C. Suppose we take a mole of pure methanol, add a mole of water to it, and heat the mixture to 65°. Will the alcohol "boil out" of the mixture?

We can find out by checking the vapor pressure. At 65° the vapor pressure of pure methanol is 760 torr, and that of pure water is 188 torr (figures of this sort are available in handbooks). But the mole fraction of methanol in our mixture is only 0.5, so the pressure due to its vapor will be 380 torr. Similarly, the water will contribute 94 torr (if Raoult's Law holds), and the total vapor pressure will be 474 torr—not nearly high enough to boil at normal atmospheric pressure.

If a series of calculations like this is done, and the results are plotted on a standard vapor pressure curve, the graph shown in Figure 16-2 results. The graph shows that the vapor pressure of the mixture will be 760 torr when the temperature is 77°C. This means that the *boiling point* of the mixture is 77° on a day when the barometer stands at exactly 760 torr.

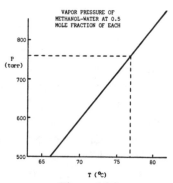

**Figure 16-2**

So far all we've succeeded in doing, after a good deal of work, is to find that an alcohol-water mixture will boil at a temperature somewhere above the boiling point of pure alcohol and below that of pure water.

But now let's look at the composition of the *vapor* in this case. Its total pressure is 760 torr, and this pressure was made up of 608 torr from the alcohol and 152 from the water. (These values could be found from the figures used to construct the graph.) These are called the "partial pressures" of the two components. If Avogadro's Principle holds, and the pressures of the gases depend on the number of molecules of each in the mixture*, then 608/760 of the molecules in the mixture must be alcohol, and 152/760 must be water. That is, the mole fractions must be about 0.8 of alcohol to 0.2 of water. If, in this distillation process, we condense the vapor, the resulting liquid will be much richer in alcohol than the one we started with: we have stepped up the mole fraction of alcohol from 0.5 to 0.8.

If we repeat the distillation—that is, take the first liquid that condenses and redistill it—the new vapor will again have a higher proportion of methanol in it. And if the process were repeated over and over again, the distillate would be practically pure methanol. This is precisely what is done when any material, from gasoline to spearmint oil, is refined by fractional distillation.

## Complications

That all sounds so straightforward that I hesitate to point out the complications. Still, the ingenuity that we'll consider shortly was called forth by the complications.

In the first place, I assumed throughout the foregoing discussion that Raoult's Law does apply to mixtures of the sort we worked with. But I have frequently said that this sort of law applies well only to *dilute* solutions. Fifty-fifty molar mixtures are far from dilute! There *are* cases where the theoretical treatment I've used is pretty satisfactory. It would work well for a benzene-hexane mixture, and it's not bad for the methanol-water example that I used. But often van der Waals forces and other factors enter in an important way—enough to make the calculations work out very badly.

For instance, in the distillation of ethanol (ordinary ethyl alcohol) from its water solutions, when the alcohol concentration reaches about 95%, the vapor in equilibrium with the solution is *not* richer in alcohol. It has the same molar concentration as the liquid. Pure ethanol has a slightly *higher* boiling point than this mixture, and repeated distillation will produce no purer alcohol. This is why most laboratory alcohol is labeled 95%. More complicated methods must be used to produce "absolute alcohol."

---

*This is called *Dalton's* law of partial pressures, and was used by Avogadro as one justification for his great hypothesis.

The second complication in fractional distillation is inherent in the process itself, even if Raoult's Law were followed. I have shown here that the vapor that is *in equilibrium* with the liquid at any time is richer in the low-boiling component—methanol in our case. But in practical distillation, this vapor is removed, leaving the remaining liquid somewhat poorer in methanol. If we start the distillation with a mixture containing 50 mole-% of methanol, it will not be long before this fraction drops to 40%, as the richer distillate is collected. Then a new set of calculations would show that while the vapor is still richer in methanol than the liquid, it has less than the 80 mole-% of the first distillate. And the longer we distill, the lower the proportion of methanol in the distillate.

### Distilling Columns

The oldest and simplest way of dealing with this complication is simply to distill until *most* of the desired ingredient has come over in the distillate, and throw away what is left in the boiling pot. With methanol-water mixtures, for instance, it works reasonably well to distill until half the liquid has come over and dump out the rest. This can be repeated, redistilling the distillate, and again discarding the last half. But this is wasteful—so equipment has been devised for *continuous, repeated* distillations.

Figure 16-3 depicts one piece of apparatus of this type. This one has the advantage of being easy to understand, though its efficiency is not high. There are two things to keep in mind as we contemplate this contraption: (a) the vapor is always richer in alcohol than the liquid it comes from; and (b) the richer in alcohol, the lower the boiling point.

We start boiling the pot at the bottom, and vapor begins to condense in the chambers A, B, and C, with fresh vapor constantly bubbling through each pool of liquid past the cone-shaped valves. In bubbler C, this liquid will be richer in alcohol than the liquid in the pot (since it is a "first distillate"). Therefore, the vapor coming from the pot will be hotter than the boiling point of the liquid in C. This incoming vapor can, therefore, make liquid C boil, while the loss of heat will make the vapor itself condense. The same thing happens again in chamber B, and again in A. Thus there are four successive distillations, though heat was applied only once, at the bottom. The bypass tubes constantly return the "spent" liquid

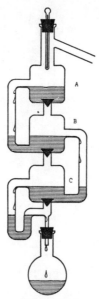

Figure 16-3

to a next lower chamber. This is the liquid that would have been thrown away in the primitive type of redistillation that I mentioned first.

From the earlier theoretical discussion, it should be clear that a "perfect" fractional distillation is one in which the very first vapor that appears is collected and redistilled. An apparatus of the sort just shown begins to approach this ideal; and if several hundred chambers were used, instead of only three, it would be possible to get practically pure methanol at the top, leaving only water at the bottom.

There are all kinds of variants of this apparatus, but surprisingly enough, one of the best is very much simpler in design. It is the "packed column," and consists simply of a straight tube (usually of glass, in the laboratory) that is filled with loose-fitting solid material. See Figure 16-4. Glass beads can be used, but other shapes are more efficient. The principle is basically the same as with the column I've already shown. Vapor condenses *on the surface* of each bead in the column, to make a thin film of liquid, which is then redistilled by the hotter vapor coming up from below. Thus, the surface of each bead performs the function of the bubbler chambers of my first column.

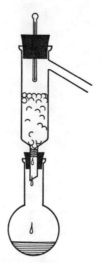

**Figure 16-4**

The efficiency of a column of this sort can be expressed in terms of "theoretical plates." A particular packed column might deliver distillate of the concentration that would be obtained by five successive "perfect" redistillations; then it would have an efficiency of five theoretical plates.

Column distillation is used industrially to separate gasoline from kerosene, diesel oil, lubricating oils, and all the other petroleum products. It is also perfectly possible to do fractional distillation at very low pressures (cf Chapter 2 on low-pressure boiling). This is done with vitamins, perfumes, and many other valuable products, some of which would be damaged if distilled at their "normal" (i.e., 1 atmosphere) boiling points.

*Ion-Exchange Processes*

Fractional distillation with a packed column may have been the first place where a basically inefficient process was made practical by repetition in small, automatic steps. Nowadays, there are so many "column" processes that we can take only a sampling here.

Ion-exchange is one such process. As usual, we'll begin with a specific case. There are many solid materials, both natural and synthetic, whose crystal lattice contains positive ions. But in some the ions are held so weakly that exchange of one positive ion for another is easy when the solid is dropped into a water solution containing an ionic solute. One such solid is a mineral of the zeolite class. If sodium zeolite is dropped into a solution containing calcium chloride, some of the sodium ions are lost, and are replaced by calcium ions. See Figure 16-5. In a given case, 60-80% of the sodium may be lost, and an equivalent amount of calcium taken from the solution. Equilibrium is finally attained, with both types of ions escaping from, and returning to, the solid.

Figure 16-5

Zeolite (or a synthetic equivalent) is used for the treatment of water for the many industrial processes which require water almost completely free of calcium. This is a requirement of laundries, steam-plants, and any number of others. Most natural water contains some calcium ions, and in some areas the water is so "hard" that soap cannot be effectively used.

To remove calcium ions, the sodium form of zeolite is packed into a column, and the exchange of calcium for sodium is permitted to occur serially. Suppose that in a given column, 80% of the dissolved calcium were removed as water passed through the top centimeter of the zeolite. Then, in the next centimeter 80% of the calcium that remained would be removed. See Figure 16-6.

Figure 16-6

Each time, the lost calcium is replaced by its ionic equivalent of sodium. A little figuring shows that under these hypothetical conditions, only a ten-millionth of the original calcium ion would remain in the water after it had passed through 7 centimeters of the column.

A column of this sort will, of course, eventually become "spent." That top centimeter will soon become practically saturated with calcium ions, having lost all its sodium, and so on down the line. But even now the zeolite is not thrown away.

We've been working with the equilibrium:

$$Na_2 Ze + Ca^{++} \rightleftharpoons 2Na^+ + CaZe \text{ (using Ze as a conve-}$$

nient symbol for the complex zeolite lattice). And I have said that the process was not very efficient. At equilibrium the reaction decidedly favors the right side of the equation, but considerable sodium ion is constantly returning to the zeolite. This fact is used in the "regeneration" of the column.

To regenerate, the column is disconnected from the process, and a concentrated solution of salt, NaCl, is passed through it. Although this reverse process is *chemically* inefficient, the column design, together with the high concentration of sodium ions in the brine, overcomes the inefficiency, and the calcium ions are "exchanged out," until the column is again full of sodium zeolite. The excess salt solution, with its calcium ions, is flushed out with clear water, and the column is ready for reuse.

These water-softening columns are familiar household items throughout the Central United States. Commercial firms install them, usually a pair of columns to a house. (They look like hot water tanks.) One column removes calcium (and magnesium, iron, etc.) ions from the "hard" water until it becomes exhausted, whereupon the householder merely turns a valve to use the other tank. The used column is picked up by the supplier and regenerated at a central plant.

The zeolite minerals are the oldest of the ion-exchange materials, but of recent years they have been largely replaced by synthetics that do the same job, often much better—that is, with much greater ionic capacities and greater versatility. Also, there are resins that will exchange negative ions rather than positive ones.

Most of the ion-exchange resins will accept hydronium ions (if they are positive-ion exchangers) or hydroxide ions (if negative). If water is passed through resins of these types in succession, all ions can be removed, except the hydronium and hydroxide of pure water itself. This "deionized" water is now often used in laboratories in place of distilled water.

Illustrated in Figure 16-7 is one such column pair that is easy to make. I used one for years, before small laboratory rigs, with "throwaway" resin mixtures, became available. I have labeled the columns to show what happens when water containing sodium chloride passes through. A is an "acid" resin, capable of existing in the hydrogen form, and B is a basic one that can hold hydroxide ions. The water emerging from tube A is weakly acidic, but after it has passed through B it is perfectly neutral. When the columns become spent, they are regenerated separately, by flushing A with hydrochloric acid solution and B with sodium hydroxide solution. The siphon tubes in the set-up illustrated are used merely to keep the resins always moist, and therefore preserve even flow.

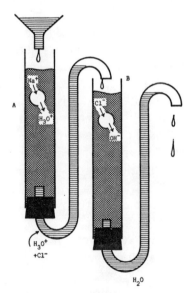

**Figure 16-7**

Ion-exchange columns can remove all charged particles in solution, and are therefore even more effective than all but the most highly efficient stills when water for electrical studies is required—so-called conductivity water. But such columns will not remove uncharged impurities, such as sugars, and organic coloring matters found in peaty ponds.

Natural ion-exchangers play an enormously important part in agriculture. Most clays in soil, and all peats, are weak exchangers for positive ions. They can hold the vitally important ammonium ions, as well as those of the trace metals, without which plant growth is impossible. All of these ions would be dissolved out of the soil by rainfall without the action of the ion-exchangers.

## Chromatography

Chromatography was invented in 1906 by Michael Tswett, a Russian botanist. His word "chromatogram" was descriptive: *chroma* means color in Greek. When certain colored materials were dissolved in a specific solvent, and then the solution was allowed to flow through a column packed with a white powder, it was seen that different pigments moved down the column at different speeds, making a "picture" (hence the "gram"), with bands of color. This can be demonstrated by one of the first uses made of the process—the separation of plant pigments.

If any fresh green plant material (grass, alfalfa, spinach) is gently dried in a cool oven, and then extracted with a solvent such as chloroform, a green solution results. The green color conceals a mixture of many different pigments, and separation of these by any of the older "classical" methods is excessively difficult.

For the chromatograph column, many different materials can be used. Tswett first used calcium carbonate. I have found calcium phosphate effective (I happened to have some on the shelf). There is almost no limit to the possibilities—you may use anything that works. The choice of solvents for a particular chromatogram can be a very delicate one, but the following crude demonstration is effective.

The lower end of a broken burette makes a good small-diameter chromatograph tube. A plug of cotton or glass wool is pushed to the bottom, and then powder is packed into the tube. See Figure 16-8. This can be done by tamping the dry powder in, a little at a time. I prefer to make a slurry of it with petroleum ether and pour this into the tube. Under the influence of gravity alone, the liquid will slowly drain from the tube, leaving a uniform and fairly solid column of wet powder. It is important not to let this column get dry at the top, or it becomes less uniform and gives poor results.

If the original chloroform extract were poured on top of this column and allowed to percolate through, the result would be

Figure 16-8

disappointing. The green solution would simply leak through, coming out almost unchanged—perhaps leaving a little color on the powder of the column. But if the extract is first evaporated to dryness to remove the chloroform (preferably under a vacuum, in a round-bottom flask or large test tube: see Chapter 2), and the residual material is dissolved in petroleum ether*, the resulting solution can be chromatographed. See Figure 16-9.

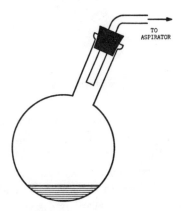

Figure 16-9

*"Petroleum ether" is a misnomer: it is not an ether, but a mixture of alkanes. Products are available with different boiling ranges. A mixture consisting mostly of heptane and boiling around 60°-70° works adequately for this demonstration.

A few milliliters of the solution are poured carefully on top of the column, with care not to disturb the powder any more than necessary. As the solution sinks down, it is followed by more petroleum ether. A narrow green band will appear at the top of the column. This color is now "developed" by adding continuously a mixture that contains 10 volumes of petroleum ether to one part of chloroform.

At first the colored band at the top widens. Then it separates into a series of bands of different colors that move down the column at different speeds. See Figure 16-10. There may be a pair of yellow bands, one or more green ones, and perhaps a clear green or blue. When these bands are well separated, the contents of the column may be pushed from the tube, and the separate pigment bands simply sliced apart, to be extracted separately.*

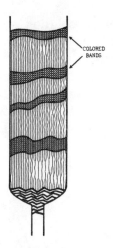

COLORED
BANDS

**Figure 16-10**

The colored bands are formed by processes similar to those of an ion-exchange column, though the specific mechanism in a given case may be less well understood. A pigment is adsorbed relatively inefficiently on the granules of solid in the column, and equilibrium is established between adsorbed pigment and dissolved pigment. Where the concentration of dissolved pigment is high, or its solubility is low, relatively much is adsorbed on the solid. This

---

*For a much more detailed and delicate series of experiments with chromatography of this and other types, see *Scientific American,* March 1969: Vol 220, No. 3, p. 124.

produces the original colored band at the top of the column. When the column is "developed" with the slightly better solvent, containing a little chloroform, pigment is eluted from the upper layer, and readsorbed, somewhat less strongly, further down. If the elution and and adsorption occur at different rates for different pigments, there will be separation. The more weakly adsorbed pigments will travel down the column more quickly, the strongly adsorbed ones will tend to stay where they are.

## Other Chromatographic Methods

As the science—and art—of chromatography progressed, the "chrom" part of it became increasingly irrelevant. Color is still sometimes used for identification, but there are other means for detecting a moving band. For instance, it may be colorless but show fluorescence in ultraviolet light. It may be radioactive, and detectable with a radiation counter.

Or there may be no attempt at all to spot the bands of material while they are on the column. Instead, the development can be carried on until one band after another is washed all the way through the column, to be collected as it emerges. Elaborate mechanical "fraction collectors" are now routinely used for this kind of work. Each of a series of dozens or hundreds of tubes will collect exactly 20 drops or exactly 2 ml. Each fraction can be analyzed (often mechanically) by measuring some property—spectroscopic, pH, or whatever.

One of the materials that can be used for packing a column is cellulose. After it was found that this worked well for conventional chromatograms, it was a short step to go to the use of ordinary filter paper *instead* of a column. A simple test of this technique is easy to do. Put a tiny spot of ball-point pen ink (or felt-tip pen ink) about 1 cm from the end of a narrow strip of filter paper, and drop the strip into a test tube that has a little solvent at the bottom. See Figure 16-11. Many solvents are effective, for example isobutyl alcohol saturated with water and containing a little acetic acid. As the solvent creeps up the paper, different constituents of the ink may move at different speeds.

As I said, the choice of solvents is critical for good work in chromatography. For a particular job, mixtures of four or more solvents may be used. Moreover, these must often be of high

**Figure 16-11**

purity (or their own impurities may "chromatograph out"); and chemical supply houses now advertise specially refined grades of solvents for chromatography.

Filter paper chromatography, that started from humble beginnings like the experiment I've just described, has now reached a high degree of sophistication. For an example of this, consider the following technique.

A specific protein under investigation is hydrolyzed to yield a mixture of many different amino acids—the "building blocks" of proteins. The whole mixture may now be treated to convert all of the acids into derivatives especially suitable for chromatography. The mixture is "spotted" on the lower left-hand corner of a *square* sheet of filter paper. See Figure 16-12. The paper is then

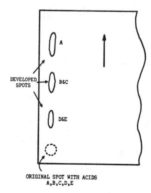

**Figure 16-12**

developed vertically in one solvent complex. The result will be a series of spots, invisible as yet, up one side of the paper. But each of these spots may still contain two or three different amino acids, all of which happened to move at nearly the same rate in this solvent mixture.

The paper is now dried, turned on its side, and inserted in a new solvent mixture, for development in a direction at right angles to the first series. See Figure 16-13. Since the solvent is different, rates may be different for certain amino acids which originally moved together, and eventually a two-dimensional chromatogram results. As the last step in the process, this pattern is made visible by spraying the sheet with ninhydrin in acetone and heating it. This treatment gives a purple color to spots of amino acid. The acids can be identified by comparison with spots of known composition.

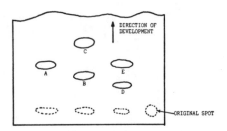

**Figure 16-13**

Within the last few years, "Thin-Layer Chromatography" has begun to take over the field for biological work of certain kinds. For this, the "column" is a very thin layer of adsorbent deposited on a piece of glass or plastic. This is treated like the filter paper that I have described, but the adsorbent can be very uniform in thickness and composition, and very tiny amounts of material— thousandths of a milligram—can be analyzed.

At the same time as these techniques were being developed, another direction was opened up when it was found that constituents of a mixture of gases can be separated by chromatography on a column that consists of an inert solid moistened with a non-volatile liquid. The column is usually in a long, rather thin tube. When a sample of mixed gases is forced into the column, the gases dissolve in the liquid that moistens the granules. Thus a "band" is formed. Now the chromatogram is developed with an inert gas like hydrogen or helium, and the sample moves through the column, its constituents being washed out by the carrier gas at one point, and redissolved further along.

In gas chromatography, no attempt is made to locate the bands of material while they are on the column. Instead, each is detected as it emerges, and the detection methods are as delicate as modern science can make them. As with thin-layer chromatography, milligrams are often considered much too big to work with, and it is common to speak of micrograms and nanograms—millionths and billionths of a gram.

# Appendix

## EXPERIMENT: THE COLORIMETRIC MEASUREMENT OF AN EQUILIBRIUM CONSTANT*

Chemical reactions occur so as to approach a state of chemical equilibrium. The equilibrium state can be characterized by specifying its equilibrium constant. In this experiment, you will determine the value of the equilibrium constant for the reaction:

$Fe^{+3} + SCN^- \rightleftharpoons FeSCN^{++}$, for which the equilibrium expression is:

$$\frac{[FeSCN^{++}]}{[Fe^{+3}][SCN^-]} = K.$$

In order to find the value of K, it is necessary to determine the concentration of each of the species $Fe^{+3}$, $SCN^-$, and $FeSCN^{++}$ in the system at equilibrium. This will be done colorimetrically, taking advantage of the fact that the $FeSCN^{++}$ ion is the only highly colored species in the solution.

The color intensity of a solution depends on the concentration of the colored species and the depth of solution viewed. Thus, e.g., 2 cm of a 0.1M solution of a colored species appears to have the same color intensity as 1 cm of a 0.2M solution. Consequently, if the depths of two solutions of unequal concentration are chosen so that the solutions appear equally colored, then the ratio of the concentrations is simply the inverse of the ratio of the two depths. It should be noted that this procedure permits only a comparison between concentrations. It does not give an absolute value of

---

*Taken, with permission, from *Experimental Chemistry*, 2nd Ed., by M.J. Sienko and R.A. Plane. McGraw-Hill Book Co., Inc., 1961. Very slightly revised.

either one of the concentrations. To know absolute values, it is necessary to compare with a standard of known concentration.

A standard solution in which the concentration of $FeSCN^{++}$ is known can be prepared by starting with a small known concentration of $SCN^-$ and adding such a large excess of $Fe^{+3}$ that essentially all the $SCN^-$ is converted to $FeSCN^{++}$. Under these conditions, you can assume that the final concentration of $FeSCN^{++}$ is equal to the initial concentration of $SCN^-$.

## Procedure

Put six clean test tubes in a rack. To the first, which is to serve as the standard, add 5 ml (precisely with a pipette) of 0.20 M ferric nitrate. For the other five tubes, proceed as follows: Measure (pipette) 10 ml of 0.20 M ferric nitrate into a graduated cylinder, add water to 25 ml, and mix thoroughly. Pipette the following amounts of this diluted solution into the five test tubes: (1) 5 ml; (2) 2 ml; (3) 0.8 ml; (4) 0.32 ml; (5) 0.13 ml. With a pipette, add water to each of tubes 2 to 5, to make the volume in each just 5 ml. You now have a series of tubes, in which the concentration of $Fe^{+3}$ varies by a factor of 0.4 from one to the next.

If preferred, there is a more accurate, if slightly more cumbersome, way of preparing the five diluted solutions. Ten ml of the original 0.20 M Fe solution is put in a graduate and diluted to 25 ml. Five ml of this solution are put in test tube (1) and then 10 ml of the remainder are diluted to 25 ml. Again, 5 ml are used in test tube (2) and 10 ml diluted, etc.

Finally add 5 ml of 0.0020 M NaSCN (or KSCN) solution to each of the six tubes. Now the problem is to determine the concentration of $FeSCN^{++}$ in each test tube relative to that in the standard tube. Take the two tubes to be compared, hold them side by side, and wrap a strip of paper around both. See Figure 1. Look down through the solutions toward a white paper on your bench. If color intensities look identical, record this fact. If not, take the standard tube and pour out some of its solution into a *clean, dry* beaker or flask (you may need to pour some back) until the color intensities appear identical. Measure the heights in the two tubes being compared in millimeters. Make this comparison for each of the five tubes.

**Figure 1**

| *Data* | Height of<br>liquid | Comparison<br>height of<br>standard liquid |
|---|---|---|
| Test tube 1 | _____ | _____ |
| Test tube 2 | _____ | _____ |
| Test tube 3 | _____ | _____ |
| Test tube 4 | _____ | _____ |
| Test tube 5 | _____ | _____ |

*Results*

| | Init. concs.<br>$[Fe^{+3}]$ $[SCN^-]$ | | Equil. concs.<br>$[FeSCN^{++}]$ $[Fe^{+3}]$ $[SCN^-]$ | | |
|---|---|---|---|---|---|
| Standard | ____ ____ | | ____ | ____ | ____ |
| Test tube 1 | ____ ____ | | ____ | ____ | ____ |
| Test tube 2 | ____ ____ | | ____ | ____ | ____ |
| etc. | | | | | |

In calculating the initial concentrations, assume that $Fe(NO_3)_3$ and the thiocyanate are each completely dissociated. Remember also that *mixing two solutions dilutes both of them.*

In calculating the equilibrium concentrations, assume that in the standard tube all the initial $SCN^-$ has been converted to $FeSCN^{++}$ (this assumption will be checked in item 1 of the discussion, below). For the other test tubes, calculate $[FeSCN^{++}]$ from the ratio of heights in the color comparison. Equilibrium concentrations of $Fe^{+3}$ and $SCN^-$ are obtained by subtracting $FeSCN^{++}$ formed from the initial $Fe^{+3}$ and $SCN^-$.

For each of test tubes 1 to 5 calculate the value of K, showing at least one such calculation in detail. Decide which of these values is probably the most reliable; or, if many of them agree reasonably well, take the mean of these.

## Discussion

1. Using the best value of K, calculate what the concentration of $SCN^-$ must have been in the standard tube at equilibrium. Was the assumption reasonably accurate that all the $SCN^-$ had been converted to $FeSCN^{++}$?

2. Why are the values of K determined by test tubes 1 and 5 probably not so reliable as the others?

3. As you have noticed, a small change in height does not appreciably affect the color intensity. Assume that in test tube 3 your height measured for the standard was 5% too great. On this basis, recalculate the value of K. Find how big a percentage error in K would have resulted from this 5% error in measurement.

# Index